HIGH TEMPERATURE LUBRICATION

High temperature lubrication

by

A. R. Lansdown

Mechanical Engineering Publications Limited
London

First published 1994

ISBN 0 85298 897 4

A CIP catalogue record for this book is available from the British Library

Printed and bound by CPI Antony Rowe, Eastbourne

CONTENTS

ABOUT THE AUTHOR

Dr A. R. Lansdown trained originally as a chemist, and received his first degree at University College, Cardiff, and his doctorate at Cambridge. From Cambridge he went to Canada as a National Research Council Post-Doctorate Fellow.

In Canada he became interested in the petroleum industry and in 1956 joined the Production Research and Technical Service Laboratory of Imperial Oil Limited in Calgary. Here he became involved for the first time in the application of chemistry to various aspects of engineering, especially the flow of fluids through the rock formations of petroleum reservoirs.

In 1961 he returned to Britain to join the Ministry of Aviation, with responsibilities for aircraft lubricants as well as hydraulic fluids and fuels. In this work he was able to combine his technical experience with his life-long interest in aviation. Like many other lubricant technologists at that time, he soon realised that the purely chemical aspects of lubricant formulation and behaviour could not be divorced from the engineering aspects of lubrication, the realisation which was soon afterwards to lead to the concept of tribology as a unifying discipline.

His period in the Ministry of Aviation coincided with the intensive search for high-temperature lubricants and other products for use in supersonic aircraft, and he became particularly interested in synthetic and solid lubricants.

In 1968 he was appointed Manager, and subsequently Director, of the newly-formed Swansea Tribology Centre, and retained this post until his retirement in 1988. In addition to the management of the Centre, he also carried out a wide variety of consultancy projects, including a number of plant failure investigations, and these further involved him in analysing the engineering aspects of the failures.

He took an active part in the activities of the Institute of Petroleum and the Tribology Group of the Institution of Mechanical Engineers. He was a member of the Council of the Institute of Petroleum from 1984 to 1990. He was awarded the Tribology Silver Medal in 1986 and was appointed a Fellow of the Institution of Mechanical Engineers in 1993.

Since his retirement from the Swansea Tribology Centre he has continued to work as an independent consultant.

FOREWORD

This book is based on a study of high-temperature lubrication carried out by the author between June 1991 and April 1992. During the period 1986–1988 the Research and Development Sub-Committee of the Institution of Mechanical Engineers Tribology Group Committee had carried out a survey of engineers to determine what aspects of tribology were considered to need the highest priority in either research or improved information. One of the topics highlighted by that survey was the need for better availability of information about high-temperature lubrication.

The Research and Development Sub-Committee decided that this need could best be met by the commissioning of a review of the subject, and the author was invited to consider writing such a review. The study was made possible by the award by the Institution of a Senior Clayton Fellowship.

To ensure that the information contained in the review was as accurate and up-to-date as possible, visits were made in the course of the study to the three major petroleum company research centres in the United Kingdom, those of British Petroleum, Exxon and Shell. The help of members of the staff of those three centres is gratefully acknowledged.

In addition, visits were made to three organizations in the United States active in the development of high-temperature lubricants or lubrication. These were the Fuels and Lubrication Division of the Wright Laboratory at Wright Patterson Air Force Base, the Lewis Research Center of the National Aeronautics and Space Administration (NASA), and the US Army Tank-Automotive Command (TACOM). Thanks are due to the US Army Research, Development and Standardization Group (UK) and the USAF European Office of Aerospace Research and Development for financial and administrative assistance in making these visits possible.

In December 1989 the author presented a paper on High Temperature Lubrication to a meeting of the Tribology Group of the Institution of Mechanical Engineers, and the paper was subsequently published in *Proceedings of the Institution of Mechanical Engineers, Part C: Journal of Mechanical Engineering Science*. Several of the ideas contained in that paper have been expanded and incorporated here.

CHAPTER ONE

Introduction

Lubrication is an ancient art and a modern science. The oldest pictorial evidence of deliberate lubrication is over 3500 years old, and the first written records are over 2000 years old.

By Roman times vegetable oils, animal fats, mineral oils, and solids were all in use as lubricants, and between them they provided excellent lubrication performance. Their high temperature capabilities were, in fact, so far ahead of man's needs that it was 1800 years before men even took any interest in establishing what their temperature limits were. In 1788 Scheele showed that the 'plumbago' which had been used for 2000 years as, among other things, a solid lubricant, was in fact two distinct substances, graphite and molybdenum disulphide, having quite different high temperature behaviour.

As a result of the Industrial Revolution, there was a rapid recognition of the temperature limitations of the common vegetable and animal oils and fats, which oxidized and hardened rapidly at high temperatures. This was a major reason for the increase in the use of the mineral oils which were becoming readily available. The mineral oils performed much better at higher temperatures, and rapidly replaced many of the natural oils, in spite of their inferior boundary lubricating ability.

It was only in the middle of the Twentieth century that a deliberate large-scale attempt was made to improve the temperature capabilities of lubricants. Even in the major conference on lubrication and wear organized by the Institution of Mechanical Engineers in London in 1937 there was little or no discussion of the temperature limits of lubricants, although there were papers about flash temperatures, oxidation, and the effect of temperature on viscosity.

The next thirty years brought great progress in high-temperature lubrication. At the 1967 Institution of Mechanical Engineers Conference on Lubrication and Wear there were several papers which referred to the temperature limits of lubricants. The figures published showed considerable advances in mineral and synthetic lubricants compared with those available in 1937. The main driving force behind this great progress was the requirements of the aircraft and space industries, including the development of successful gas turbine engines. These requirements had led to considerable sums of money being spent on research on high-temperature lubricants.

At one end of the scale the most promising organic structures, such as polyphenyl ethers, fully fluorinated hydrocarbons, and perfluoropolyethers, had been synthesized, and their performance and limitations had been well studied. At the other end of the scale the capabilities of mineral oils had been raised to a peak in the deep dewaxed super-refined (DDSR) mineral oils developed by E. E. Klaus and his co-workers.

However, by 1967 there were already doubts as to whether the rate of progress, or indeed any progress, could be maintained in the development of improved organic compounds for high temperatures. These doubts were, in fact, well founded. The progress since 1967 has been far less than in the previous thirty years, and a list of present-day high-temperature liquid lubricants would differ little from those described at the 1967 conference.

Some people would claim that this lack of progress has been due to the much smaller sums of money spent on research on high-temperature lubricants since 1967, but this is probably not true. It is certainly true that much less money has been spent than previously, but the governments of the United States and the Soviet Union, among others, were quite prepared to provide funds for promising projects. The problem has been that few people have successfully identified useful lines of research.

The greatest progress since 1967 has been in the perfluoropolyalkylethers (PFAE) and the synthetic hydrocarbons, but in both cases the progress has been steady and not spectacular.

Several United States organizations, especially the United States Air Force and the National Aeronautics and Space Administration, have carried out or sponsored investigations into the perfluoropolyalkylethers, and the performance and properties of these fluids now appear to be well established. They seem to be the best liquid lubricants for use at very high temperatures up to about 400°C, and their main limitations are poor boundary lubrication and extremely high cost.

The practical temperature limits for synthetic hydrocarbons are much lower, with lives of only a few hours at 200°C. Their initial thermal degradation temperatures are much higher than this, but they oxidize readily, and this determines their practical limits.

The above figures illustrate one of the difficulties in discussing high-temperature lubrication, namely the problem of defining what is a high temperature. In the context of grease lubrication, high-temperature problems can arise at any temperature over 130°C. For lubrication of vehicle engines with mineral oils, sump temperatures over 150°C are probably too high, but cylinder temperatures will be over 200°C. Solid lubrication of ceramics can be successful up to 1000°C.

It follows that for the engineers and technologists for whom this book is written the definition of high temperature cannot be made in terms of a specific temperature; any temperature high enough to cause problems in a particular situation is a high temperature. As a working definition, for the purposes of this book, high temperature will, therefore, be taken to mean any temperature high enough to cause problems which would not arise at lower temperatures.

Figure 1 shows the temperature–pressure envelopes for several different fields of application of lubricants. This figure was published in its original form by Beerbower and Appeldorn in 1967 and would in most respects still be relevant twenty-five years later. The major change is represented by the discontinuous line which has now been added to show roughly the boundaries that would apply to the more advanced low-heat-rejection engines.

It can be seen that the temperatures envisaged for low-heat-rejection engines are no higher than those which were required for certain other applications in 1967, but the nature of the lubrication problem differs in several important respects. The lubrication techniques which are used in hot metal working or for glass moulding would require radically different design approaches to enable them to be used in internal combustion engines. The highest temperatures shown for aerospace and aviation would have involved solid lubricants of various types, and the problems in applying them to engines are very far from being solved.

In other words, it is not simply the temperature which limits the usefulness of lubricants. Lubricants have existed for many years which can survive the actual temperatures needed by present-day equipment. It is the nature of the mechanisms, the required life, the quality of lubrication required, or some other aspect which makes lubricants unusable for particular applications.

Really effective lubrication is only available at temperatures below about 150–170°C. At higher temperatures compromises are always necessary in terms of reduced life, lower loads, lower sliding speed, higher levels of friction, difficulties of lubricant feed, avoidance of compatibility problems, or other constraints.

The function of a lubricant is to reduce or control friction, but where

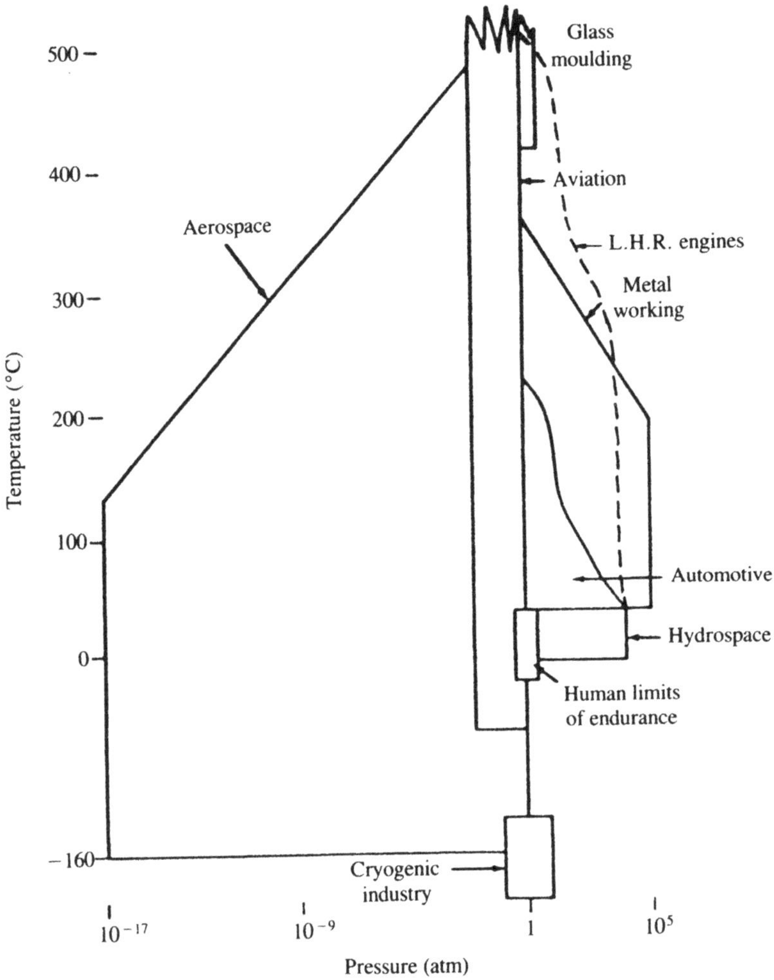

Fig. 1 The physical lubrication environment *(Reproduced by permission of World Petroleum Congresses)*

compromises must be made it would be a dangerous over-simplification to say that the objective is to achieve minimum friction. A better definition of the overall target is to minimize the power loss involved in achieving and maintaining relative movement between load-bearing surfaces. Negligible or even zero friction can be achieved in gas bearings and magnetic suspensions, but for carrying significant loads power must be dissipated in supplying external gas pressure, or an electricity supply for electromagnets, so that these techniques are not particularly efficient.

If we consider only more conventional systems, coefficients of friction as low as 0.001 can be achieved in rolling bearings and oil-lubricated plain bearings. Grease-lubricated rolling bearings have coefficients of friction between 0.005 and 0.02, while the best solid lubricants may achieve 0.02. All these values represent a power wastage of 2 per cent or less, and by any standards that is a reasonably low figure.

Where performance compromises are necessary to provide lubrication at high temperature, it may become important to decide what levels of friction are acceptable. Any choice of a target coefficient of friction may be arbitrary, and the problem needs to be considered in relation to the design and performance of the whole system. Some workers have suggested that a criterion of good lubrication is a coefficient of friction of 0.1 or less, but this represents a power wastage of 10 per cent, which is clearly not very efficient. The values of 0.2 or higher which are claimed for the coefficient of friction of some high-temperature solid lubricants are, in fact, very wasteful.

It might be argued that any step forward is of value, but this is not necessarily true, and the case of low-heat-rejection engines provides an illustration of the conflict.

Conventional automotive engines have very low overall fuel efficiencies, generally between 20 and 35 per cent. Much of the loss of efficiency is due to the need to cool the engine, and significant reductions in fuel consumption could be achieved if artificial cooling could be eliminated. The object of the considerable volume of research on low-heat-rejection engines is to improve fuel efficiency by perhaps 20–25 per cent. It follows that if the higher temperatures can only be achieved by the use of lubricants giving coefficients of friction of 0.2 or more, then the frictional power losses may offset any gain in efficiency from the higher operating temperatures.

It must be emphasized that this review is concerned only with lubrication, in other words the use of materials to reduce or control friction. The subjects of unlubricated sliding, wear, and bearing materials are not primary subjects for discussion.

Lubrication is not directly related to wear. As a rule, lubrication tends to reduce wear, but this is not always the case, and situations exist in which wear is increased in the presence of lubricants. There are also techniques for reducing wear which do not reduce friction, and may increase it. The use of hard surface coatings is an obvious example of such techniques, which are wear-reducing but not lubricating. Wear, or wear-resistance, is therefore not considered in this review, except where it influences lubrication, as in the case of solid lubricants.

The subject of bearing materials is too big to be considered at length, except where the composition of the surface influences the behaviour of the lubricant. Traditionally lubricants have been developed for use in contact with metals. The metals widely used have included steels, cast irons, copper alloys, and white metals, which comprise alloys and composites of lead, tin, or aluminium. The detailed information on lubricant behaviour in Chapters Seven–Nine refers, generally, to use with metals, unless otherwise specified.

The behaviour of lubricants in contact with ceramics has been found to differ in some ways from their behaviour with ferrous metals, and differences are likely to be found with some of the higher-temperature and refractory alloys. For this reason, the lubrication of ceramics will be discussed separately in Chapter Eleven.

Lubricants, especially solid lubricants, may be incorporated in the structure of a composite bearing material, and the subject of self-lubricating composites will be considered in Chapter Nine.

As a result of interest in low-heat-rejection engines, a considerable amount of research is being done at present on some aspects of high temperature lubrication. Because of this, new materials and lubrication techniques are likely to be developed, and some of the specific temperature limits quoted in later sections of the review will eventually become obsolete. However, the extent of improvement during the period 1970–1990 has not been great, and present indications are that major improvements in the temperature capabilities of lubricants will be very hard to achieve.

In situations in which the temperatures present are approaching the limits either for lubricant performance or for lubricant life, some traditional assumptions are no longer adequate. For example, it can often be assumed that the temperature of a bearing is equal to the outlet temperature of the oil from the bearing assembly. In critical temperature conditions it must be recognized that there is no such thing as a single bearing temperature, and that the highest temperatures present may have no resemblance to the outlet temperature.

As a general rule the upper temperature limits for lubricants are lower than those for the materials of construction in a system, so that the maximum operating temperature of many systems is imposed by the limitations of the lubricants. One exception is the elastomeric material used in seals and hoses, which have temperature limitations comparable with those of lubricants.

The aim of this book is to provide information which is of value to all engineers faced with problems of lubrication at high temperatures, whether those temperatures are at less than 140°C or more than 1000°C. In order to achieve this, the various sections fall into two different categories. In the first category is information about the general factors which affect lubrication as the temperature rises. The second category attempts to define the actual temperature limits for the use of specific types of lubricant.

In both categories the objective has been to provide the information in a simple form which will be easily understood by non-specialists. At times this attempt will have failed, while at others it may have resulted in over-simplification, and a specialist may identify apparent distortions or inaccuracies. Subjects such as oxidation, thermal degradation, and boundary lubrication could be described in far greater detail than is possible or desirable in this review.

The intention has been to describe each topic in sufficient detail to enable an engineer to understand the factors involved in solving a problem of lubrication at high temperature, but without unnecessary complication. For those who would like to follow up any particular topic more thoroughly, a short bibliography of suggested further reading is provided.

Those who are looking for the answer to a specific question may find it by going directly to one of the later Chapters Seven–Thirteen, but for an adequate understanding of the problems involved, reference to the earlier chapters is recommended.

CHAPTER TWO

Factors Affecting High Temperature Lubrication

It cannot be emphasized too strongly that, in many cases, high temperatures in lubrication are created unnecessarily by the user, and in a few cases even by the designer. The following are common examples of incorrect operation which can cause overheating.

Operating too fast.
Operating too slowly.
Operating at excessive power.
Allowing the oil level to fall.
Supplying too much oil.
Filling a bearing with grease.
Inadequate cooling.
Inadequate running in.
Operating continuously equipment designed for intermittent operation.

The problem of over-greasing a bearing will be described in Chapter Eight, but the same problem of continuous churning can cause overheating even with oil. One popular make of car some years ago suffered from engine overheating problems because the recommended oil fill caused sump temperatures to rise to 160°C. Reduction of the oil fill by a quart brought the temperature down by 20°C.

The type of use can lead to overheating problems. Failure to follow the recommended running-in procedures in some types of delivery van led to temperatures over 200°C in their rear axles. London taxis experience unusually high temperatures up to 150°C in automatic transmissions because of the frequent gear changes caused by long periods of driving in heavy traffic. Similarly, some models of car in police service have suffered from high gearbox temperatures because of the nature of their duty.

It is often possible to get away with a great deal of improper lubrication or improper operation, but when overheating does occur, especially in a mass-produced item, it will always be valuable to look for any faulty procedures. In the words of the old joke 'When all else fails, try reading the instruction manual'.

The effects of high temperature on lubricants may be physical or chemical. Physical effects are usually, but not always, reversible, and include melting, vaporization, changes in adsorption, phase changes, and thermal expansion, which in turn causes changes in density and viscosity. Small but significant changes also take place in properties such as specific heat, thermal conductivity, surface tension, and the pressure coefficient of viscosity. Chemical changes are usually, but not always, irreversible and include oxidation, thermal degradation, chemisorption, and reaction with other materials, especially bearing materials and elastomers.

Chemical changes are time-dependent, so that, for example, the extent of deterioration due to oxidation or thermal degradation depends not only on the temperature of the lubricant but on the time of exposure to that temperature. Physical changes with temperature are normally assumed to be instantaneous, but this is probably an over-simplification, and hysteresis effects may be significant in the short time-scale of asperity contacts. Temperature changes in a lubricant are, themselves, not instantaneous, but take place at rates which are determined by the flow rate, the specific heat, and the heat flux both into and within the lubricant.

Where lubrication problems exist at high temperature, it is obviously important to analyse the overall system in order to make use of all possible approaches to minimize the thermal stress on the lubricant. To achieve this it is necessary to analyse the nature of the lubricated system and the nature of the various heat sources, and to understand the various effects of heat on the properties and performance of the lubricant.

The most important effects of heat on a lubricant are viscosity effects, oxidation, and thermal degradation, and each of these is the subject of a later chapter. Several effects, such as adsorption, chemisorption, and chemical reaction, are important in boundary lubrication, which is also the subject of a later chapter. The remaining influences will now be discussed in this chapter.

2.1 RESIDENCE TIME

It will be seen later that the time for which a lubricant is exposed to high temperature is of vital importance in defining the maximum temperature for use of a particular lubricant. Conversely, an important factor in designing a system for high temperature operation is to ensure that the lubricant is not exposed to high temperatures long enough to suffer unacceptable deterioration.

There are many obvious examples which illustrate this point. Mineral oils used as automotive engine lubricants will experience transient temperatures well over 200°C between the top piston ring and the cylinder liner. If the sump temperatures were as high as this the long residence time would result in the oil becoming unusable in a matter of hours. In the same way, a temperature which would be high for long-term survival of the lubricant in an automatic lighthouse or navigation buoy would be trivially low for the short operating life of a guided missile.

Unfortunately it is not easy to define the residence time of a lubricant at a particular temperature, because lubricant temperatures vary in both time and place even in the simplest lubricated systems.

It is conventional to speak about the temperature of a lubricated system, but there is of course no such temperature. All lubricated systems experience power dissipation, and the dissipated energy provides a heat flow. This heat flow is concentrated at specific locations, including, particularly, those where the greatest friction force is present in moving contacts. Thus, heat is inevitably supplied to the lubricant at discrete non-uniform points in the system and the temperature of the system and of the lubricant varies as a result.

Where the lubricant is fixed in the same place in a component, as with solid lubricants and often with greases, the temperature will be the greatest where the greatest frictional power dissipation is taking place, and there will be a strong thermal gradient through the volume of lubricant. Definition of an effective lubricant temperature, and consequently of a residence time at a particular temperature, will therefore be difficult.

Where lubricant replenishment or recirculation is taking place the problem is further complicated by the fact that an element of lubricant passes through the system, gaining heat in some places and losing it in others. The processes of gaining and losing heat are themselves time-dependent, so that, as pointed out earlier, there will be a series of gradual increases and decreases of temperature.

Figure 2 shows a hypothetical plot of temperature against time for a volume of oil passing from a reservoir by means of a pump into a working component such as a gearbox and returning to the reservoir. In such a simple case it would obviously be possible to calculate an average oil temperature for the whole system, and thus establish a residence time for

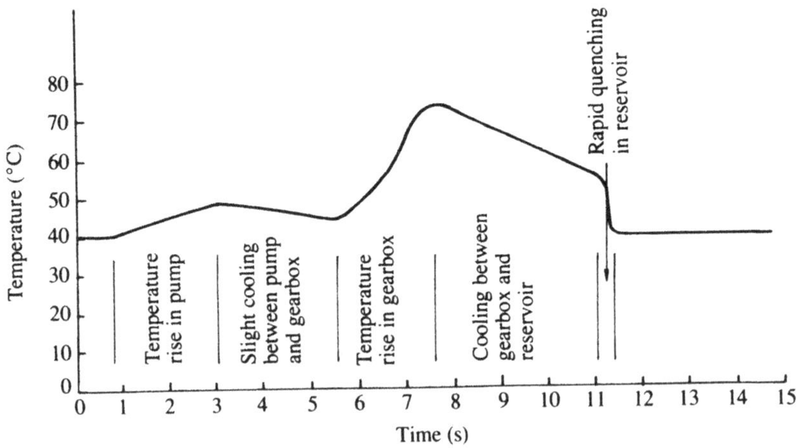

Fig. 2 Hypothetical temperature/time plot for oil passing through a gearbox

the oil at that temperature. Unfortunately, such an approach would give a serious underestimate of the lubricant deterioration because deterioration does not vary linearly with temperature.

The deterioration of a lubricant takes place by a variety of chemical reactions, some of which will be described in detail later. Each reaction has its own rate, which is dependent on the activation energy, so that no accurate general relationship can be used to relate overall deterioration to temperature.

For the purposes of this discussion a general relationship can be used, which states that the rate of a chemical reaction doubles for every 10°C rise in temperature. This is a rule-of-thumb which has been used by chemists for many years, although it is recognized to be a gross oversimplification. Based on this rule, we can use the plot in Fig. 2 to calculate relative lubricant decomposition rates, and these are plotted against time in Fig. 3.

It can be seen that, at the peak temperature reached at the outlet from the gearbox, which is only 34°C higher than the reservoir temperature, the deterioration of the lubricant is over ten times as rapid. In practice the relationship between rate of decomposition and temperature is more complicated than is indicated by the above rule-of-thumb. For thermal decomposition an approximate relationship can be established experimentally, but for oxidation no useful correlation can be established,

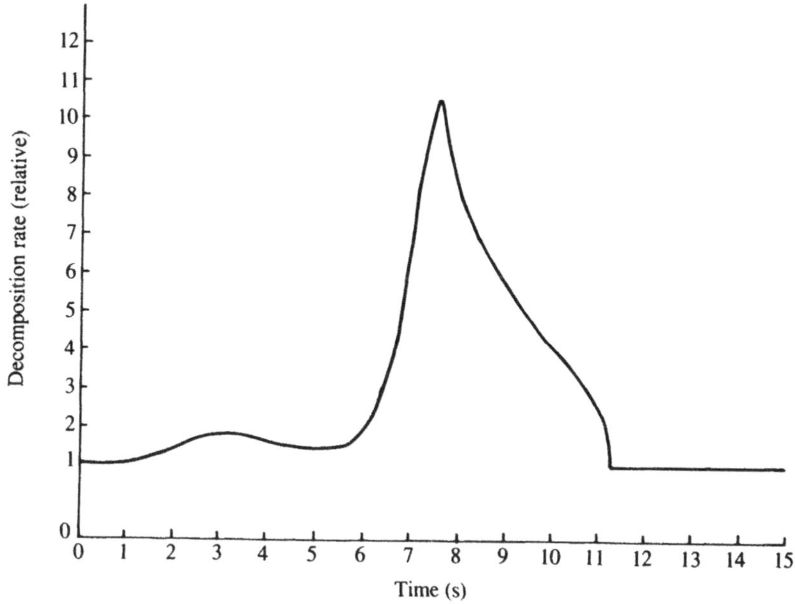

Fig. 3 Variation of deterioration rate of lubricant from Fig. 2

because the availability of oxygen and the degree of mixing of air and lubricant will strongly affect the rate of oxidation.

Figure 2 emphasizes the importance of residence time, especially at higher temperatures. Except in the most complex systems, it is usually possible to make useful estimates of residence time in various components. At the design stage, or even later, it is possible to reduce the effective residence time in the hottest zones by increasing the volume of lubricant or its flow rate, or by dividing the flow so that a single element of lubricant is not made to pass through a series of hot zones.

2.2 TEMPERATURE PROFILES

In order to work out the residence time of a lubricant at different temperatures in a machine or plant, it is necessary to establish what are the temperatures at various points in the machine. In an existing machine, sufficiently accurate profiles can often be measured by means of thermocouples, pyrometers, or other thermometric techniques. At the design or modification stage it will be necessary to analyse the system and assess the sources and magnitude of the different heat inputs. This should then make it possible to evaluate the temperature profiles, and to establish the periods for which each volume of lubricant is exposed to the various temperatures in the system. The lubrication system can then be designed to minimize the overall thermal stress on the lubricant.

Even in a simple lubricated system there will be several different sources of heat input. Overall, the variety of heat sources is very wide, but it may be useful to divide them all into four categories which have different influences on the designer or the operator. These four categories are ambient temperatures, machine or system inputs, component inputs, and flash temperatures.

2.2.1 Ambient temperatures

The ambient temperature may be defined as the temperature of the environment in which the machine or system has to operate. It is outside the control of the designer or the manufacturer of the machine, although they should always be aware of the proposed ambient temperatures and take them into account. It may be outside the control of the user of a machine, although the user always needs to be aware of the ambient temperature, so that he can take any appropriate action to mitigate its worst effects.

In many cases the ambient temperature can be controlled, at least to some extent, by the user of equipment, because the environmental temperature is not simply the atmospheric temperature. It may be considerably modified by the characteristics of the building in which the equipment is housed. In temperate, sub-arctic, or arctic geographical areas modern buildings are generally designed to ensure the comfort of the occupants, and the resulting temperatures will be reasonable for most mechanical equipment. Even so, equipment failures have often been caused by over-heating due to direct sunlight on a dark roof, even in sub-arctic areas. In tropical or sub-tropical regions the problem is usually to provide adequate ventilation.

The ambient temperature will also be modified by heat from the machines themselves or from other manufacturing processes. In some industries such heating can dwarf the direct atmospheric temperature, although again, in modern industry, the need to provide an acceptable working environment for the work-force will often ensure an adequate ambient temperature for the equipment. In the case of heat generated by equipment and processes, the resulting ambient temperature for a particular machine is, to a much greater extent, under the control of the operator, who can ensure adequate space between units, and locate critical equipment so that it is not exposed to unnecessary heat.

The extremes of atmospheric temperature throughout the world range from −80°C in some parts of Antarctica to +55°C in the Sahara. The annual variation in temperature in a particular location may also cause operating problems, and this is as high as 95°C in Siberia, although in most places it is much lower, being only about 35°C in most maritime areas. More information about world climatic temperatures can be found in the *Tribology Handbook* (Butterworth, London, 1973) and the *Handbook of Aviation Meteorology* (HMSO, London, 1960).

In general ambient temperatures have the advantages of being easily measured, relatively constant, and fairly predictable.

Heat usually flows from the lubricated system to its environment because power is always dissipated in a lubricated system, resulting in a net generation of heat. The ambient temperature is important because it controls the cooling of the machine. The rate of heat loss will be directly related to the difference between the effective ambient temperature and the effective system temperature. As a first approximation it can be assumed that this difference will remain constant, so that an increase in ambient temperature will produce a roughly equal increase in system temperature.

2.2.2 Machine or system heat inputs

Any lubricant in use will be subject to a variety of heat inputs from different parts of the lubricated machine or lubrication system itself. There will also be some locations in which heat is lost, or in other words the heat input is negative.

Overall, for a lubrication system which is in an equilibrium state of operation, the loss of heat must be the same as the gain of heat, but it is important for two reasons to know the net increase in temperature at any point in the system. The first reason is that the effectiveness of lubrication in any component is highly dependent on the temperature of the lubricant at that point. The second reason is that, as shown previously, the deterioration of a lubricant is much more a function of the highest temperatures than of the average temperature.

It is also true, of course, that even at equilibrium the temperature at any point will be directly related to the various heat inputs. If there is an increase in heat input, the temperature will continue to rise until the rate of heat loss becomes equal to the heat gain, and a new higher equilibrium temperature is reached.

System heat inputs will arise from many different sources such as fuel combustion, electrical power, work done on products, chemical processes, and friction from bearings, gears, valves, seals, pistons, cams, and so on.

Where lubrication is performed by a small quantity of lubricant retained in place in a component without any recirculation, the heat inputs will reach the lubricant by conduction through the moving parts or the supporting structure. There will also be heat generated by frictional power dissipation within the lubricated component, but this aspect is considered in the next section. Heat will also be lost through the moving parts or the supporting structure. Even in this relatively simple situation it can be valuable to assess the heat sources because of their influence on thermal gradients and on the equilibrium temperature of the lubricant.

A similar situation exists where there is a facility for re-supply of lubricant to a component, but without recirculation, as with centralized grease or oil lubrication or with oil mist or fog. Such systems are often used in situations where there is severe thermal stressing of the lubricant, resulting in rapid lubricant degradation. Analysis of the various heat inputs can result in a reduction of the overall heating of the lubricant, and permit economies by the use of lower lubricant feed rates, as well as the possibility of avoiding over-heating problems. Reducing the lubricant feed rate can be particularly valuable where product contamination is very undesirable, such as in paper, textile or food manufacture.

The most complex mechanical systems, and those where there is a high level of power dissipation, are usually lubricated by means of oil circulation. Because of their complexity, such machines often contain a large number of heat inputs of different types. A volume of oil passing through the system will be subjected to a series of temperature changes as it passes the different heat sources.

If we consider a volume of the oil passing through a hot zone, and assume that at the point of entry and the point of exit its temperature is uniform, then the outlet temperature will be related to the inlet temperature by an equation in the form of equation (1)

$$T_{\omega} = T_{\alpha} + t_1 C_1 (T_1 - T_{\alpha}) \tag{1}$$

where T_{α}, T_1, and T_{ω} are the inlet, zone, and outlet temperatures, t_1 is the residence time in the zone, and C_1 is a heat-transfer coefficient.

Within the hot zone the oil temperature will, of course, not be uniform, either axially or laterally. Lateral differences in temperature will arise due to the differences in distance from the heated surfaces, modified by the degree of mixing taking place. In the boundary layer adjacent to a hot surface a small element of the oil will be exposed continuously to the surface temperature and will be more extensively degraded than the bulk of the oil.

In an extreme case the oil in the boundary layer may be completely degraded and produce a coating of solid breakdown products or even carbon on the surfaces. The new solid surface will itself create a boundary layer in the flowing oil which will break down in turn, causing a thickening of the layer of breakdown products. The layer of degraded lubricant will usually have a lower thermal conductivity than the original surface of the vessel, so that the deposition process usually ends when, because of the insulating effect of the deposits, the adjacent oil is no longer exposed to temperatures high enough to cause deposits of breakdown products to form.

Even where the surface temperatures are not high enough for deposits to form, the oil close to hot surfaces will be hotter than the bulk, and will therefore be degraded more severely than would be anticipated from the average oil temperature in the zone, or even from the outlet oil temperature.

In the absence of any significant cooling, the outlet oil temperature from the first heat source will be the inlet temperature for the second, and so on. From equation (1), for a series of heated zones,

$$\begin{aligned} T_{\omega} = T_{\alpha} &+ t_1 C_1 (T_1 - T_{\alpha})(1 - t_2 C_2)(1 - t_3 C_3) \ldots (1 - t_n C_n) \\ &+ t_2 C_2 (T_2 - T_{\alpha})(1 - t_3 C_3) \ldots (1 - t_n C_n) \\ &+ t_3 C_3 (T_3 - T_{\alpha}) \ldots (1 - t_n C_n) \\ &+ \ldots \\ &+ t_n C_n (T_n - T_{\alpha}) \end{aligned} \tag{2}$$

The actual calculation might be possible for some simple systems, but the value of equation (2) lies more in the indications it gives of the effects of a series of heat sources. Bearing in mind that each of the functions $t_n C_n$ must be positive and less than unity, inspection of equation (1) provides support for some useful subjective ideas about lubricant temperatures.

(a) The importance of the initial inlet temperature diminishes as the complexity of the oil flow path increases, but never disappears.
(b) The final zone temperature is the most important in determining the final outlet temperature.
(c) The shorter the residence time of the oil in hot zones, the lower will be the temperature rises.
(d) Insulation of hot zones will reduce the heat transfer coefficients C_n and thus reduce the temperature rises.
(e) It can be seen that if any zone temperature is less than the initial inlet temperature of the oil there will be a comparable cooling effect. Expansion of the equation confirms the obvious assumption that any zone which is cooler than the oil inlet temperature to that zone will also produce a cooling effect.

It is often assumed that the outlet temperature of the oil from the last zone is equal to the temperature of that zone, and that deductions about the oil life can be made on the basis of the final outlet temperature. It can be seen that this is true only if the heat transfer in the last zone is perfect, in which case $t_n C_n$ in equation (2) becomes equal to unity and T_ω becomes equal to T_n. In most real situations T_ω will be less than T_n, and it follows that some part of the oil will have been subjected to temperatures greater than T_ω. The overall oil degradation will then be greater than it would be at a uniform temperature of T_ω and the oil life will be correspondingly less.

It is therefore important, particularly in high-temperature systems, to recognize that the outlet temperature will normally be lower than the highest temperatures experienced by the oil.

The final outlet temperature of the oil, or the temperature of the hottest zone, may also be quoted as the operating temperature of the system. It is important for the optimization of a high-temperature system to recognize that there is no such temperature as the operating temperature of a system. There will always be a series of operating temperatures, each of which should be separately assessed, empirically if not quantitatively.

One final point about system heat inputs is that the overall temperature rise can be reduced by reducing the number of heated zones traversed by a single volume of lubricant, or, in other words, by providing a number of separate parallel flow paths. Alternatively, a supply of oil at the initial inlet temperature can be arranged to bypass some of the earlier hot zones and combine with the heated oil at a later point in the system. Such techniques will add to the complication of the system, but may be justified where system temperatures are high enough to cause lubrication problems.

The overall heat balance, and the inlet temperature, are determined by the total heat inputs and by the cooling available in the reservoir or coolers. It is when the capacity for cooling the oil is limited that the greatest problems arise.

2.2.3 Lubricated component heat inputs

The discussion in the previous section is generally applicable to all types of heat source which affect the lubricant. This section now considers the special case of lubricated components in which the source of heat is frictional heating of moving contacts.

In such components the heat generation is confined to a very small part of the lubricant and of the total surface area of the bearing assembly. As a result that small element is exposed to much higher temperatures than the bulk of the lubricant. Thus, whereas with other types of heat source it may be justifiable to consider only the total heat input and temperature rise, this is not adequate for consideration of a bearing assembly or other tribological component.

Where full fluid film separation of the bearing surfaces exists, the power loss is entirely due to viscous friction in the lubricant film in the loaded zone, and the heat generation takes place entirely in the lubricant film. Where the lubricant film is too thin to maintain full fluid film separation, the friction will be partly viscous and partly surface friction due to asperity interactions between the opposing surfaces. The heat generation will be partly in the lubricant film and partly concentrated at the interacting asperities.

The concentration of part of the power dissipation at isolated asperities leads to very high 'flash temperatures' at those asperities, and these are discussed in the next section. Some aspects of the special problems of greases and solid lubricants will also be considered in their separate chapters.

For completeness it should be mentioned that where the bearing or gear, or a flange or flinger ring, is partially submerged in a pool of lubricant, there will also be some power dissipation where the component passes through the lubricant, but in a properly-designed and lubricated assembly, this will be small compared with the power loss in the loaded zone, and the resulting heat will be dispersed through a much larger volume of lubricant. Any resulting temperature rise will be small.

In all cases the greater part of the heat will be generated within, or in contact with, a thin film of lubricant. As a result that small quantity of lubricant will experience the highest temperature in the assembly. Even within the lubricant film the heat generation is not uniform, but is highest at the point of closest approach of the bearing surfaces. To assess the effect on both the performance and the life of the lubricant it is necessary to try to estimate those high film temperatures.

The problem of estimating the temperature variations in the bearing assembly differs considerably depending on whether the assembly is self-contained or fed by oil circulation.

Self-contained bearings will be discussed first. These are bearings which have a small quantity of oil contained in the bearing assembly without any facility for oil circulation or cooling. They have great advantages in simplicity and cost, but for high-temperature applications they have the major disadvantage that heat can only be removed by conduction through the shafts and housings, and eventually by radiation from external surfaces.

Professor Dudley Fuller has suggested that for simple forms of self-contained bearing with mineral oil lubrication the maximum film temperature should not exceed about 82°C, and that the limiting shaft diameter and speed to remain within that limit are as shown in Fig. 4. A shaft speed higher than 4000 r/min will almost always be too high for a self-contained bearing. Higher temperatures will be acceptable with certain synthetic oils, and information on the oil life at different temperatures for a number of different oils is given in Chapter Seven. In all cases, when synthetic lubricants are used to withstand higher operating temperatures, care must be taken to ensure that the bearing materials are also suitable for the higher temperatures.

The heat generated in the bearing will be removed by conduction along the shaft and by radiation and convection from the housing.

The efficiency of heat removal from the bearing oil film into the bulk of the oil and thence to the housing depends on the type of lubrication

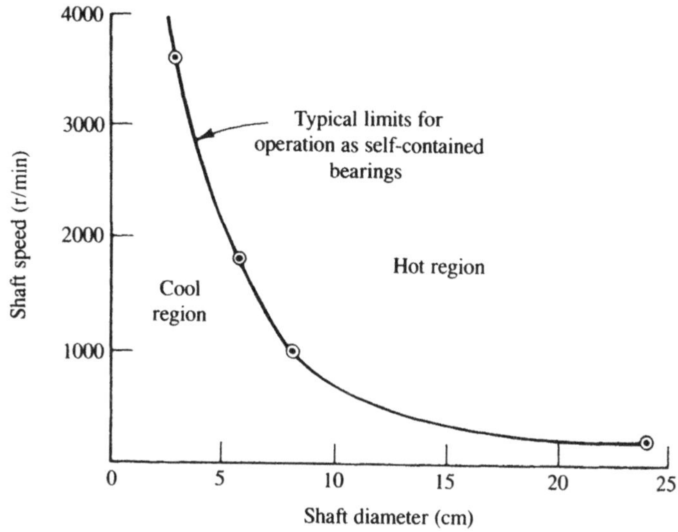

Fig. 4 Typical limits for self-contained bearings *(From D. D. Fuller,* Theory and practice of lubrication for engineers, *1984. Reproduced by permission of John Wiley and Sons)*

system which is used. The best cooling is obtained where the bearing is partly submerged in an oil bath, or where a flinger ring or disc rotates through the oil and throws the oil over the bearing and journal. A simpler and less effective method uses a drip feed, a wick-feed or impregnated cotton waste to feed oil to the journal, but these are not likely to be suitable for any application where high-temperature problems occur.

Figure 5, also by Fuller, shows the temperature difference which is likely to exist between the bearing oil film and the inner wall of the housing for the three different types of oil feed. The spread within each sector depends on whether the outer surface of the housing is losing heat only by radiation into still air, or whether air flow over the housing helps to give convective heat loss.

The proportion of the total heat lost along the shaft will generally be much lower than through the external surfaces of the housing, but there are many factors which influence this. Obviously any external heating on the outside of the housing, or any shielding or insulation, will reduce the ability of the housing to disperse heat. Conversely, if the outer surface is exposed to an airflow, the rate of heat loss can be considerably increased, so that fan cooling can be very effective.

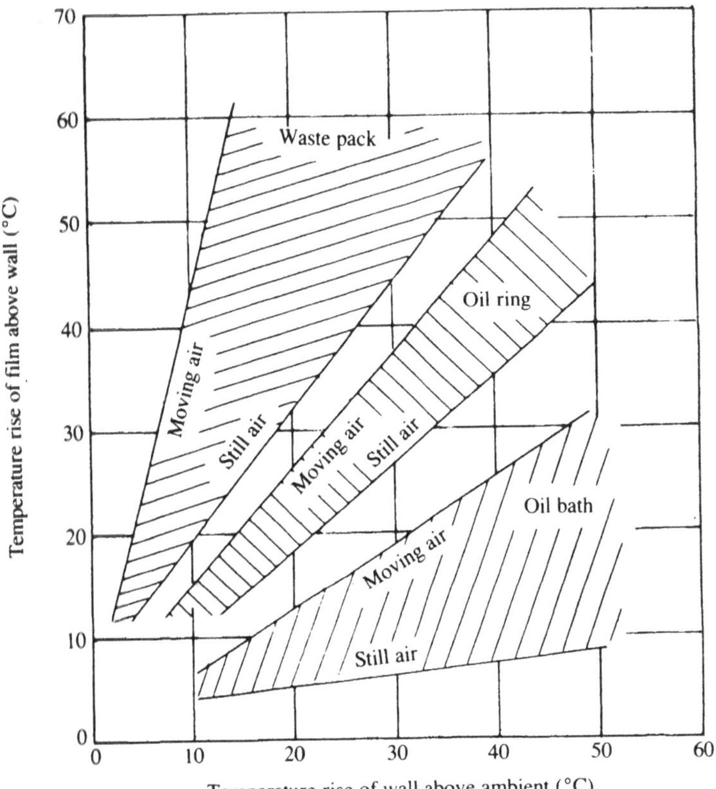

Fig. 5 Temperature gradients in self-contained bearings *(From D. D. Fuller,* Theory and practice of lubrication for engineers, *1984. Reproduced by permission of John Wiley and Sons)*

The ability of the housing to disperse more heat than the shaft is partly a function of the greater surface available to conduct or radiate heat. Heat loss can be increased by the provision of external fins on the housing, as is used routinely for automotive air-cooled engines.

In any type of bearing the moving parts are usually smaller than the stationary parts, and as a result heat loss is usually greater through the stationary components. There are exceptions, such as tilting-pad, thrust, and journal bearings, where the moving surfaces are larger than the stationary surfaces, and the latter can have a tendency to overheat.

Using some simple assumptions and iterative approaches it is possible to calculate the heat balance in a simple bearing assembly. Some procedures are contained in the books listed in the bibliography.

Overall, the ability of a self-contained bearing assembly to disperse heat is limited. Where the shaft speed exceeds the limits in Fig. 4, or where the heat loss from the housing is restricted because of high ambient temperatures or lack of cooling air, there will be a risk of over-heating. A circulating oil system is then likely to be preferred, as the circulating oil has the potential to provide very effective cooling.

In order to provide oil circulation it is now normal practice to use an oil feed pump, although there are exceptions in which some feature of the machine design is used to induce the required oil flow. The use of a separate oil pump frees the designer from the constraints of using only the available oil flow, and enables him to control the position and quantity of the oil supply to meet the requirements of each component.

The calculation of the oil-feed requirements is a normal part of the design process for a bearing or gear. The result gives the minimum quantity needed to meet the lubrication requirements of the component. The minimal oil flow for lubrication purposes in a circulating system will obviously be related to the lubrication requirements of the individual components.

In practice, however, the need for oil circulation is usually determined by the requirement for cooling, and the quantity required will be calculated to ensure that film temperatures and bulk temperatures are kept within acceptable limits. The quantity of oil needed for effective cooling will commonly be much higher than that needed for lubrication.

In a properly-designed plain journal bearing with full fluid-film separation of journal and bush the film temperature should not be more than 20°C higher than the inlet temperature. The power loss in a bearing can be estimated from a diagram such as Fig. 6, which is taken from *The Tribology Handbook*. The temperature rise through the bearing will be given by equation (3)

$$\Delta T = \frac{H}{C \times Q} \tag{3}$$

where H is the power loss, C the specific heat (thermal capacity), and Q the volume flow rate.

The normal flow rate through the bearing will depend on the loading, the eccentricity, the oil viscosity, the inlet pressure, and the length/

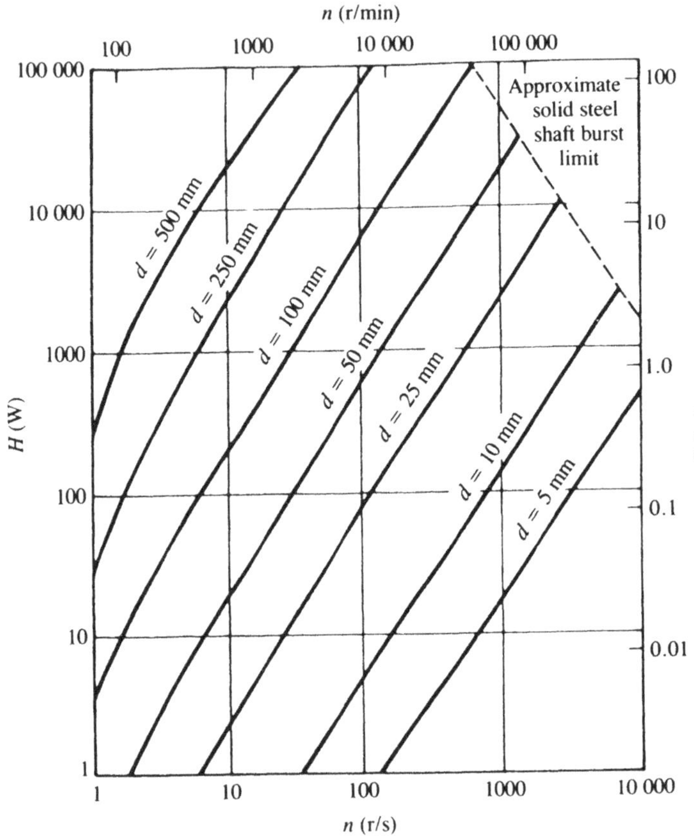

Fig. 6 Power loss from plain journal bearings *(Reproduced from* The Tribology Handbook *by permission of the editor, M. J. Neale)*

diameter ratio. A typical 25 mm diameter bearing, with length equal to diameter, rotating at 6000 r/min might have a flow rate of the order of 3×10^{-6} m^3/s and a power loss of 100 W. With an oil specific heat of 2×10^6 J/m^3.°C, this would give a temperature rise

$$\Delta T = \frac{100}{3 \times 10^{-6} \times 2 \times 10^6} = 16.7°\text{C}$$

This is the steady state temperature rise between inlet and outlet, and the peak temperature rise reached at the centre of the loaded zone will be higher, probably exceeding the recommended 20°C maximum. An acceptable approximation for the average viscosity in the oil film is based on a temperature rise 0.7–0.8 times that at the outlet.

Variations in viscosity across the lubricant film can help to increase the load-carrying capacity, so that the importance of the higher peak temperature at the centre of the loaded zone is more for its contribution to lubricant degradation than for viscosity reduction.

2.2.4 Flash temperatures

When full fluid separation is not maintained between bearing surfaces, there will be random contact between asperities on the two surfaces. Adhesive friction then takes place and the total friction increases. The additional frictional energy dissipation takes place at the discrete asperity contacts, and because of the high concentration of dissipated energy at those microscopic contact points, very high transient temperatures occur. These temperature transients are called flash temperatures.

The total energy dispersed at a single asperity contact depends on the contact stress, the coefficient of friction, and the relative speed. The temperature rise then depends on the specific heat of the materials and their thermal conductivity.

Many equations have been published for the calculation of flash temperatures, covering a wide range of speeds and contact geometries. They give results which are reasonably consistent with each other, but for metallic materials the calculated temperatures tend to be much higher than temperatures measured experimentally. One possible reason for this is that all the calculations assume that all the frictional energy is converted to heat. This is a reasonable first approximation, but since the calculations also assume that contacting asperities are elastically or plastically deformed, account needs to be taken of stored energy in elastic deformation, and the work done in creating plastic deformation.

These calculations are also of limited practical value because of some other important uncertainties. One of these is in the composition, and therefore the physical properties, of the material of the asperities. In a simple case, any practical engineering metal will have a surface film of oxide. It is possible to obtain an estimate of the effective film thickness of the oxide on a static metal surface, but during sliding this will be reduced by wear and increased by oxidation under the influence of frictional heating. It has even been suggested that oxidation can take place due to the flash temperature during an asperity contact, but this seems very unlikely because of the difficulty of oxygen access during the contact.

The thermal conductivity of a metal is of the order of ten times as high as the thermal conductivity of its oxide. The flash temperature is inversely proportional to the thermal conductivity, so that the flash temperature of an asperity of solid oxide will be ten times as high as that of unoxidized metal. In practice, of course, the effect of a thin oxide film will be less, but still significant. Other surface films, such as those formed by reactive extreme-pressure additives, will have similar effects.

Other factors which lead to uncertainty in the calculation of flash temperatures are probable hysteresis in the recovery in deformed asperities, and variations in the specific heat and thermal conductivity of elastically deformed materials or during plastic flows.

The calculation of flash temperatures in a specific situation is, therefore, complicated and imprecise, but it is very useful to have some idea of the magnitude of flash temperatures and of the contribution which they make to the overall energy balance and to lubricant degradation.

Typical flash temperatures for good-quality lubricated engineering surfaces are often quoted as 150–400°C above the adjacent bulk temperatures of metals. However, one of the earliest papers suggested flash temperatures of 600°C for lubricated metal surfaces. Where the bearings consist of low-conductivity materials such as plastics or ceramics, the flash temperatures will inevitably be higher, and this will also be true when metals are coated with non-metallic coatings for wear resistance. Flash temperatures of 1000°C or higher have been calculated for such materials.

The actual maximum temperature at the asperities is the sum of the flash temperature and the bulk temperature. In this context the bulk temperature must be taken as the adjacent surface temperature, to a depth sufficient to contain all the heat loss from the asperity during the period of the temperature flash. This surface temperature will have been influenced by the overall surface friction and by any cooling by the lubricant.

In high-temperature systems the 'bulk' surface temperatures will themselves be high. Even for well-lubricated metals a flash temperature of 200°C added to a surface temperature of 500°C leads to a very high asperity temperature, and asperity contact needs to be minimized.

There is a limit to the extent to which the surface and flash temperatures can be added together. It is recognized that the melting point of the material of the asperity imposes a practical limit on the asperity temperature. It has been found, for example, that as the relative sliding speed between two surfaces is increased, the measured asperity temperatures increase until a limit is reached beyond which no futher increase occurs. This limit may be at the normal melting point of the bearing material, but the melting point may in any case be affected by the high contact pressure at the asperity. Any other endothermic phase change taking place will also impose a limit on the asperity temperature.

The duration of the temperature flash for all practical bearing or gear systems probably lies between 10^{-8} and 10^{-4} seconds. The small amount of lubricant trapped at the asperity will therefore be subjected to very high temperatures for that short period, and a significant proportion of it is likely to decompose. This decomposition will generally be endothermic and will help to reduce the actual asperity contact temperature.

With oil lubrication the small quantity of degraded oil will rapidly be dispersed in the bulk of the oil. With grease lubrication the degraded material will be retained within a very small total grease volume, while with solid lubricants the degraded material remains in place in the surface film. The effect of flash temperatures on the life of the lubricant will therefore be considered separately later for the different types of lubricant.

The heat generated at asperity contacts contributes to the total frictional heating of the surfaces and of the lubricant film. The removal of heat from the system has been discussed in the previous section, but it should be noted that with a grease or a solid lubricant there will be no convectional heat loss. Only conductive and radiative cooling are available, and the actual thermal conductivity of greases and of most solid lubricants is very low. As a result the maximum bearing speed with grease or solid lubricants is much lower than with oils.

In any lubricated contact an increase in temperature leads to a reduction in the film thickness of either oil or grease and thus to an increase in the extent of asperity contact. This will in turn lead to an increase in frictional heating. If the system is able to disperse the additional heat which is generated, then an equilibrium will be established.

If the system is not able to disperse the heat generated at any time, the temperature will continue to rise. A new equilibrium may then be established at a higher temperature, but there may be a condition in which macroscopic hot spots occur, where larger areas reach temperatures typical of flash temperatures. Such hot spots are known to arise in brakes. If steels are present, these local hot spots may initiate the severe wear phenomenon known as scuffing and lead to seizure or tearing of the counterfaces. Scuffing will be considered in more detail in Chapter Twelve.

2.3 MELTING

The phenomenon of melting is so well known that it may seem unnecessary to consider it here, but it is so fundamental to some aspects of high temperature lubrication that this chapter would not be complete without some consideration of its effects.

The importance of the melting point (or freezing point) is that it places an upper limit on the use of solids and a lower limit on the use of liquids.

Conventional liquid lubricants freeze at temperatures which are too low to be of interest in a document about high-temperature lubrication, but some of the liquids which are potentially useful as high-temperature lubricants have melting/freezing points which are high enough to cause serious handling difficulties at normal temperatures. Their melting points are therefore of considerable importance, and will be discussed in Chapter Seven.

There are a few materials which have useful lubricant properties in both solid and liquid form. Examples are lead, block greases, certain soaps and waxes, and some thermoplastic polymers, and those which are of any importance will be discussed in Chapters Seven, Eight, and Nine.

The nature of melting varies for different types of material. A pure chemical element or compound will usually have a definite sharp melting-point at which all the material present will change from a true solid state to a true liquid state or vice versa. In such cases the physical change is from a highly ordered state of the solid to a fully disordered state of the liquid, and the solid formed on freezing is commonly crystalline. The change of state requires an input of heat, which is called the latent heat of fusion. Once melting starts, additional heat added to the material serves only to melt more of it, and the temperature remains constant until the whole mass has melted, when the temperature will again rise.

With amorphous solids the change of state is less clear-cut, and takes place over a range of temperatures. As the temperature rises the solid becomes progressively softer and there is a gradual change from plastic

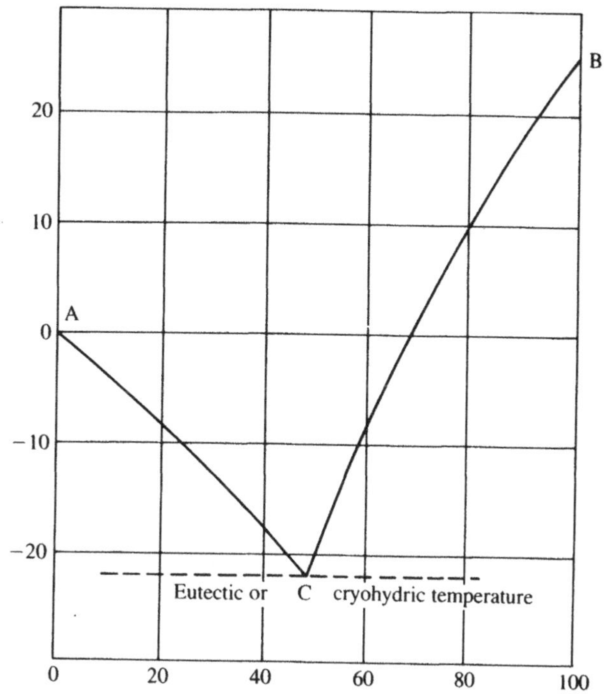

Fig. 7 Variation of melting point with composition of a solid solution

shear to viscous flow when stress is applied. There is no distinct melting point, and the definition of when it becomes liquid is subjective. At any point in the melting process, part of the heat input serves to change the state of the material and the remainder serves to increase the temperature. Although there is a latent heat involved in the change of state, it may be difficult or even impossible to measure.

A third class of material includes solid solutions and alloys. Their behaviour is easier to understand from consideration of freezing instead of melting. The situation is shown in Fig. 7. Whenever one substance (the solute) is dissolved in a second substance (the solvent) the freezing point of the solvent is depressed, and the amount of the depression increases as the concentration of the solute increases. On cooling a solution of solute A in solvent B a point is reached where some of the solvent freezes. This leaves the solution with a higher concentration of A in B, and, therefore, with a lower freezing point. Further cooling will now reach the new freezing point, where more of B freezes, and this process will continue, with more of B being progressively frozen while the solution becomes more and more concentrated.

If we now consider starting from a different case where the solution consists of a small amount of B dissolved in A, progressive cooling will result in A freezing while the solution of B in A becomes more concentrated. In either case it can be seen that at the freezing point the solution becomes saturated with the solvent, so that some of it separates out. Eventually, as either freezing point curve descends, a point is reached where the solution is saturated with respect to both components, and they both separate out together. This point is called the eutectic point, and the composition at that point is called the eutectic mixture or eutectic alloy.

In order to produce alloys having compositions different from the eutectic, a solution must be cooled rapidly so that all components solidify together.

The reverse process occurs when a small part of an alloy or solid solution melts, and unless the alloy was a eutectic the composition of the molten material will be different from that of the solid. Cooling will then cause resolidification to give a portion or layer which still has a different composition from the bulk.

A component which solidifies out of a solution is normally in crystalline form, but rapid quenching of a non-eutectic solution gives an amorphous dispersion of grains which are internally crystalline.

Since most practical metals in bearings or gears are alloys, their behaviour on melting is on the lines described above. Exceptions are surface platings such as cadmium, chromium, or silver, which are pure metals and have a single distinct melting point.

Polymers may show a fairly distinct melting point, or may show the general softening and slow melting characteristic of amorphous materials. Many polymers have another characteristic temperature at which they cease to be rigid solids and become rubbery, and this is called the glass transition temperature. As the temperature rises above the glass transition temperature they will eventually show the typical progressive softening and liquefaction of amorphous substances.

It was mentioned in the previous section that the flash temperature in an asperity contact may raise the asperity temperature to the melting point of the material, and that this may help to limit the temperature rise.

When the temperature reaches the melting point, a small portion of the asperity will melt, but heat energy is required to supply the latent heat of fusion of that portion. As a result, heat is absorbed in the molten material, and there is an interruption in the temperature rise.

The first material to melt will be at the point of maximum heat generation, so that the contact is now between a solid asperity on one surface and a film of molten material on the other. The contact stress and the coefficient of friction will, therefore, also be reduced, so that there will be an immediate reduction in the heat energy liberated. As a result, once a thin surface film has melted, there will be little tendency for more material to melt unless the initial melted film is wiped away and further solid/solid contact takes place. In this way the beginning of melting at an asperity contact helps to create a state of equilibrium, limiting the amount of damage to the asperity.

Once the asperity transit has ended, there will be rapid cooling and any remaining molten material will solidify. In most cases its composition and physical properties will no longer be identical to those of the original asperity. There may be transferred substances from the counterface and from the lubricant, and there may also have been phase or grain size changes. Such changes may be beneficial, such as a reduction in shear stress, or detrimental, such as embrittlement.

The possibility of melting at an asperity contact should therefore be considered in the selection of suitable mating materials for high-temperature bearings.

Apart from any resulting physical or compositional changes, melting at asperity contacts is likely to be beneficial in reducing asperity heights and improving running-in.

The melting process is a change from a relatively ordered solid state to a disordered liquid state. The increasing atomic mobility as the temperature rises can cause other changes which render the bearing material unsatisfactory, such as softening and lower resistance to creep. Figure 8 shows typical reduction in hardness with increasing temperature for a bearing steel, and Fig. 9 shows the reduction in creep stress with increasing temperature.

Phase changes in an alloy may also take place well below the melting point. This is particularly true for steels, and bearing steels for high temperature use need to be stabilized for the proposed operating temperature.

Ceramics have very high melting points, and studies of their physical

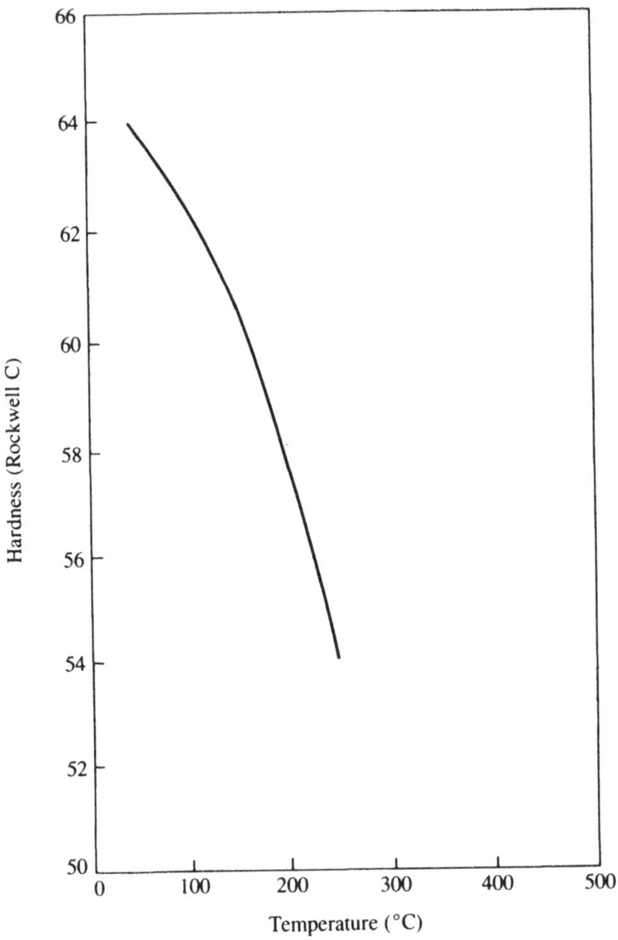

Fig. 8 Change of EN31 steel hardness with temperature

properties as they approach their melting points have so far been fairly limited. Some important chemical interactions have been found at temperatures well below the melting temperatures for some ceramics.

Some thermoplastic polymers, such as PTFE, will sinter before reaching their melting points. Thermosetting polymers will decompose instead of melting.

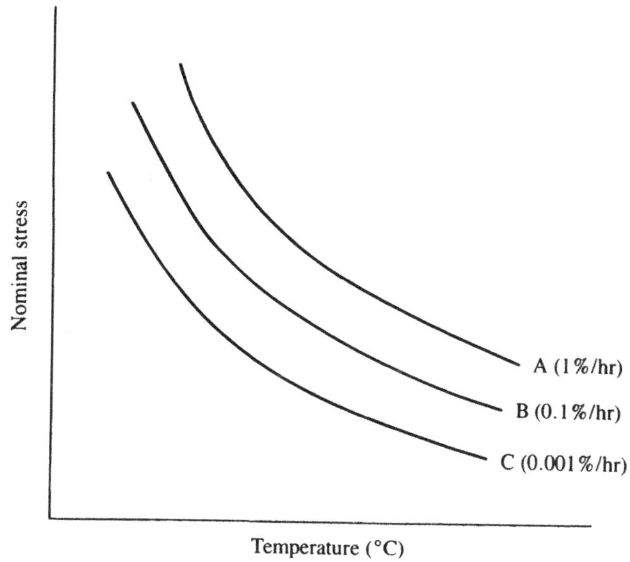

Fig. 9 Second-stage creep rate vs temperature and nominal stress *(From S. Shigley and C. Mischke,* Standard Handbook of Machine design, *1986. Reproduced by permission of McGraw-Hill)*

It follows that, although for most classes of solid material the melting point represents an absolute limit to their high-temperature capabilities, many materials will suffer some other deterioration which imposes a temperature limit well below the melting point.

Table 1 lists examples of solid materials of tribological importance with their melting points and notes on other factors which limit their high-temperature use.

2.4 THERMAL EXPANSION

The thermal expansion of materials is well recognized by engineers as important in all situations where temperatures change during operation of machinery. The problem is two-fold. First the design must accommodate dimensional changes with temperature so that parts of a structure, and particularly bearings and gears, are not overloaded. A well-known

Table 1 Melting points of some tribological materials

Material	*Melting point (°C)*	*Other changes occurring below melting points*
Metals		
Steels	1350–1450	Phase changes, reduction of hardness
Bronzes	900–1050	Lead component melts much lower
White metals	240–350	Reduction of fatigue strength
Polymers		
PTFE	325 (glass t.t)	Sinters below 300°C
Acetal	175	Softens well below melting point
Nylon	220–250	Softens below melting point
Phenolics	None	Decompose above 150°C, but usable above decomposition temperature
Solid lubricants		
Molybdenum disulphide	1185	In air oxidized above 350°C
Graphite	3340–3390 (subl.)	In air oxidizes above 500°C
Graphite fluoride		Decomposes above 470°C
Calcium fluoride	1423	Only lubricates above 250°C

example is the need to permit axial movement of one of the support bearings of a shaft which will be subject to temperature changes.

The second problem is to predict the thermal gradients which will arise in operation, so that either the design or the material selection will ensure that clearances are not lost. For example, the heat generated in a rolling bearing on a rotating shaft will generally lead to the inner race being hotter than the outer race because the housing provides a better heat sink. As a result the internal clearance may be lost and the bearing will fail.

High temperature operation can accentuate both of these problems. The greater temperature rise causes greater dimensional changes due to thermal expansion, and it is even more important to ensure that the system can accommodate those changes. The thermal gradients are also likely to be greater, and in some systems may be harder to predict because of superimposed heat from external sources.

The greater differences in coefficient of thermal expansion when using more exotic materials such as high-temperature polymers, ceramics, and refractory metals will also increase the extent of the problems. The coefficient of expansion tends to increase as the temperature rises, so that it is important in designing for high temperatures to ensure that the correct coefficients of thermal expansion are used for the whole of the relevant temperature range.

2.5 SPECIFIC HEAT

The specific heat of a substance is a measure of the amount of heat required to raise its temperature. In SI units it is the number of joules needed to raise the temperature of one kilogram by 1°K.

The importance of specific heat is that it indicates the ability of a substance to absorb heat, and in general a high specific heat is advantageous. For cooling purposes a liquid with a high specific heat is desirable, and water is the best of all coolants in this respect, having a specific heat of 4.2×10^3 J/kg°C. This compares with figures such as 2.0×10^3 J/kg°C for mineral oils and esters. The figures for metals are much lower, such as 0.5×10^3 for steels and 0.9×10^3 for aluminium, but these are on a weight basis, so that on a volume basis they would be more comparable with the figures for the liquids. A given volume of steel, therefore, provides a heat sink intermediate between that of the same volumes of mineral oil and water.

The specific heat of a material changes with temperature, but there is no consistent pattern in the variation, and for accurate calculation of thermal balances detailed tables must be used.

2.6 THERMAL CONDUCTIVITY AND DIFFUSIVITY

The thermal conductivity of a substance is a measure of the rate at which it can conduct heat from a hot zone to a cold one. It can be considered as the ratio of the rate of heat transmission to the thermal gradient, and is given by equation (4).

$$k = \frac{Ql}{A(T_1 - T_2)} \qquad (4)$$

where Q is the quantity of heat per second, l the path length, and A the area of cross-section with a temperature difference $T_1 - T_2$. In SI units it is measured in J/m°Cs or W/m°C. For the maximum removal of heat from a hot zone the thermal conductivity should be high, and for maximum insulation it should be low. Some typical figures are 94 W/m°C for iron and 1.5 W/m°C for silica.

The thermal conductivity is of value for calculating the thermal gradient between hot and cold zones where the system is in equilibrium. Until an equilibrium has been established, heat entering the material will warm up that material while it is being conducted through it. The conductor is thus providing a heat sink as well as a conductor of heat.

Under these circumstances both a high specific heat and a high thermal conductivity are useful in removing heat from a hot zone. The property of the material which combines the two factors is the thermal diffusivity. In the calculation of flash temperatures at asperity contacts, it is the diffusivity which determines the heat removal from an asperity during the short period of the transient flash.

2.7 RADIATION

When heat reaches the outer surface of a body it can be lost by conduction, radiation, or convection. Conduction in a gas is poor, so that radiation and convection become the most effective mechanisms for heat loss.

The rate of radiation from a body depends on its temperature, material, and surface condition. The effect of temperature is simple, since the rate of radiation is in proportion to the fourth power of the absolute temperature, and this relationship is shown in Fig. 10. Obviously radiation becomes much more effective at higher temperatures, and the radiation from a body at 600°C is thirty times as great as at 100°C.

The effect of material and surface condition is less simple, but can perhaps be best explained on the basis of the fact that the best reflectors are the worst radiators. This follows from the fact that a body's ability to emit radiation is exactly the same as its ability to absorb radiation. When radiation falls on a surface part is reflected and part absorbed. The more reflective the surface, the less efficient it is as an absorber of radiation and, therefore, the less efficient as a radiator.

It is common knowledge that light-coloured surfaces reflect more heat than dark ones, and that smooth polished surfaces reflect better than dull, rough ones. It follows that the best radiators are dull dark surfaces, and the ideal surface is spoken of as a black body radiator.

The properties of absorption and reflection should not be confused

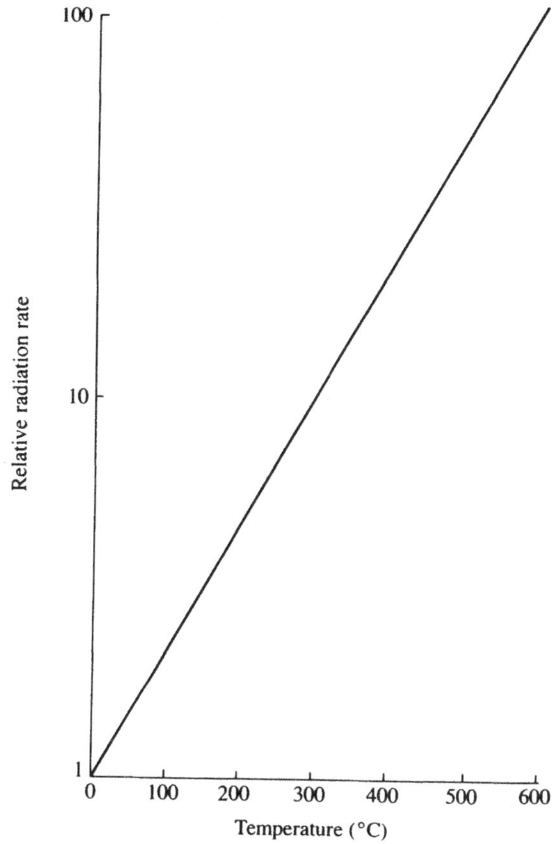

Fig. 10 Variation of radiation rate with temperature

with transmission or transparency. Glass seems to be a poor reflector of visible light because it is transparent to it, but it is also a poor absorber and radiator. The confusion is made clear if a glass surface is inclined at an acute angle to the incident light, when the reflection approaches 100 per cent.

The absorption and emission of radiation by a body are also affected to some extent by the wavelength of the radiation, so it should not be assumed that a surface which is reflective to light is equally reflective to heat.

Radiation from a surface is not influenced by the composition of a gas in contact with it. However, if a highly reflective surface is in the path of the radiation from a hot surface, a significant part of the radiation may be reflected back onto the hot surface. If this occurs, the heat radiated back onto the hot surface will be absorbed, because the absorption of radiation by a surface is not affected by radiation emitted by the same surface.

It follows that if the radiative heat loss from a surface is important, it should not be placed near to another surface which will reflect back the radiation. If, for any reason, it is essential to place the hot radiator near another body, then whenever possible the other body should be given a heat-absorbing surface so that it does not reflect back heat. Under certain circumstances it may be desirable deliberately to place a liquid-cooled heat-absorbing shield adjacent to a hot surface, where for any reason it is not possible to use liquid cooling directly on the hot surface.

2.8 CONVECTION

Convection is the transfer of heat by a fluid, liquid, or gas, in which the heat is transported by means of physical movement of heated fluid. This can usually be assumed to involve the movement of the heated fluid to a cooler location where its heat is lost either in the bulk of the fluid or to a secondary medium.

The important properties of a fluid for efficient convective cooling are low viscosity for easy flow and high specific heat. In both respects water is the best practical convective coolant, so much so that considerable efforts are made to use it at temperatures below its normal freezing point by mixing it with other substances, or above its normal boiling point by pressurization.

Where water is not suitable, many other liquids can be used. The choice will depend on the maximum temperature to which they are exposed, or simply on availability, as in the case of cooling by fuel in aircraft.

At temperatures which are too high for any practical liquid, gases can be used. They have the advantage of low viscosity, but their heat capacities on a volume basis are very low. For example the heat capacity of water at 100°C is 4.216×10^3 J/kg°C, which is equivalent to 4.04×10^6 J/m^{3}°C, while that of steam at 100°C is approximately 1.20×10^3 J/m^{3}°C, and that of air is slightly lower.

As a result gases are not very effective convective coolants, and forced convection is desirable when cooling with air or other gases.

2.9 VAPORIZATION

Almost all materials, solid or liquid, have a small amount of their vapour present adjacent to their surfaces, except when they are at 0°K (−273°C). When the vapour cannot escape from the vicinity of the surface an equilibrium is set up, vapour condensing on the surface at the same rate at which it evaporates from the surface. The partial pressure of the vapour when equilibrium exists is the vapour pressure of the substance at that temperature.

When the vapour is not constrained to remain in the vicinity of the surface it will diffuse away into the atmosphere, being replaced by fresh vapour evaporating from the surface, so that there is a gradual loss of material from the surface.

The vapour pressure of most solids at normal temperatures is very small. It only becomes important in high vacuum, when the loss of material from a solid surface may be significant and is known as outgassing. There are no materials of tribological interest which have significant vapour pressures in the solid state, and it is only the vaporization of liquids which is important.

As a liquid is heated, its vapour pressure increases with temperature until it is equal to the ambient pressure, and the liquid then boils. At the boiling point a change of state occurs in which the molecules absorb energy and increase their motion so that they lose cohesion. This absorption of energy has a similar effect to that at melting in that the addition of heat at the boiling point only increases the kinetic energy of the molecules and does not increase the temperature of the liquid or the vapour.

The temperature will remain constant until all the liquid has vaporized, and the heat absorbed by the liquid during vaporization is called the latent heat of vaporization.

Boiling is important in high temperature lubrication for two reasons. The most important is that it imposes an absolute limit on the temperature at which a liquid can be used, although often a lower practical limit is caused by oxidation or thermal degradation.

Boiling is also important because the latent heat of vaporization can be used for cooling. If a coolant in contact with a hot surface can be allowed to boil, the total heat sink due to specific heat and latent heat of vaporization will be much greater than the cooling available only from specific heat.

CHAPTER THREE

Viscosity

3.1 GENERAL

The viscometric properties of oils and greases are of paramount importance in their performance as lubricants. Taking the simplest category first, the creation of an oil film with a Newtonian oil between relatively-moving surfaces follows the generalized Reynolds equation, equation (5).

$$\frac{\partial}{\partial x}\left(\frac{\rho h^3}{\eta}\cdot\frac{\partial p}{\partial x}\right)+\frac{\partial}{\partial z}\left(\frac{\rho h^3}{\eta}\cdot\frac{\partial p}{\partial z}\right)=6(U_1-U_2)\frac{\partial(\rho h)}{\partial x}+12\rho V+12h\frac{\partial p}{\partial t} \quad (5)$$

where x and z are the coordinates along and across the film, h is the film thickness, p the pressure, U_1 and U_2 the velocities of the two surfaces along the oil film, and V the squeeze film velocity.

The only properties of the oil in the equation are the density ρ and the viscosity η. In most cases the oil is considered incompressible so that the density term disappears and the only oil property affecting the film thickness is the viscosity.

Similarly for elastohydrodynamic lubrication of nominal line or point contacts in rolling bearings or gears, the behaviour is described by the Dowson–Higginson equation or one of the subsequent derivatives. The Dowson–Higginson equation can be written for the contact of two cylinders

$$h=\frac{1.6\alpha^{0.6}(\eta_0 U)^{0.7}(E^1)^{0.03}R^{0.43}}{w^{0.13}} \quad (6)$$

where α is the pressure exponent of viscosity or pressure-viscosity coefficient, η_0 the viscosity at ambient pressure, E^1 represents a modulus parameter for the two surfaces, R is the effective combined radius of the two cylinders, and w is the load per unit length at the contact. Here again, the only properties of the lubricant are the viscosity in the entry region and the pressure exponent of viscosity.

In equations (5) and (6) the viscosity is considered to be a constant, except for the variation of viscosity with pressure which is taken into account in equation (6). Specifically, it is assumed that the temperature is constant throughout the fluid film, and that the lubricant is Newtonian, in other words that its viscosity does not vary with shear rate.

In practice, neither of these assumptions will necessarily be correct. The temperature will certainly vary through the fluid film, and the effect of temperature variation needs to be considered. Certain oils and most greases are known to be non-Newtonian, and there must be a possibility that when shear occurs in an elastohydrodynamic conjunction the behaviour of all oils or greases ceases to be Newtonian.

Both viscosity and the pressure exponent of viscosity vary with temperature. The viscosity of any liquid decreases with increase in temperature, as shown in Fig. 11, which shows the change in absolute viscosity with temperature for five different oils.

The viscosity is also often plotted in kinematic viscosity, in centistokes (cSt) or metres squared per second ($m^2 s^{-1}$). However, the absolute viscosity (centipoise, cP, or Pascal seconds, Pa s) is the product of kinematic viscosity and density, and is the property used in bearing design or film thickness calculations. Because density also decreases with increase in temperature, absolute viscosity decreases more rapidly than kinematic viscosity as the temperature increases. Kinematic viscosity, therefore, understates the effect of temperature on viscosity, and for high-temperature considerations it is particularly important to use only absolute viscosity.

The variation of viscosity is often described by the Viscosity Index (VI). This is a number which compares the change of viscosity with temperature of the oil under discussion with that of an oil from a United States Gulf Coast petroleum (VI = 0) and a United States Pennsylvania petroleum (VI = 100). Approximate viscosity indices are shown in Fig. 11 for four of the five oils.

Strictly speaking, viscosity index only applies to values between 0 and 100 and to temperatures below 100°C. Outside these ranges the figures are extrapolated, and they will still be useful probably up to about 160°C. For higher-temperature applications it is safer to use viscosity temperature curves of the type in Fig. 11.

3.2 FULL FLUID FILM LUBRICATION

The importance of viscosity in a hydrodynamic journal bearing is clearly seen in Fig. 12, in which the coefficient of friction is plotted against the

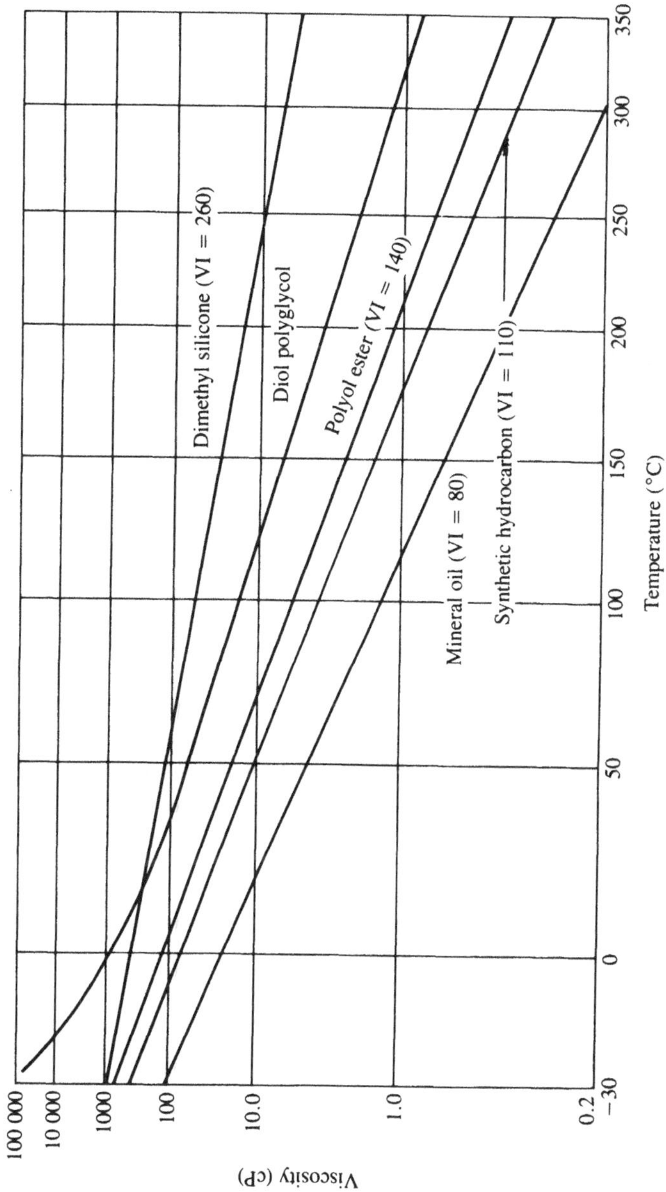

Fig. 11 Variation of viscosity with temperature

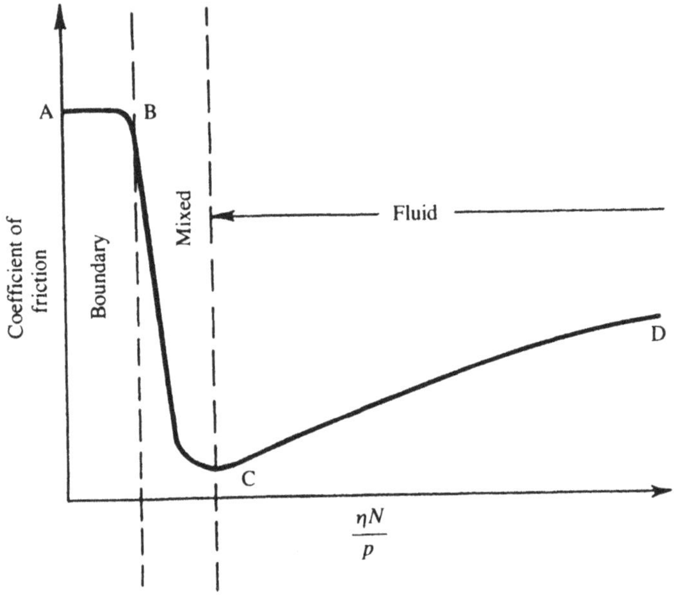

Fig. 12 Plot of coefficient of friction vs $\eta N/p$

Hersey Number, $\eta N/P$. N represents the rotational speed of the bearing, and P the specific load or bearing pressure. This type of curve is commonly called a Stribeck Curve, although it appears to have been first used in this form by McKay and McKay.

The zone CD represents the region of full fluid film separation of the bearing surfaces, and AB the region of boundary lubrication. In zone BC there is a mixture of phenomena, including some hydrodynamic separation, some boundary lubrication, and possibly some features which do not exist in either of the other zones. The features of zones AB and BC will be considered in Chapter Six.

If in zone CD the bearing speed and load are held constant, the coefficient of friction depends only on the oil viscosity, and the whole of the friction is due to viscous shear. At high viscosity the shear stress is high, but as the viscosity decreases the coefficient of friction falls until the point C is approached. This decrease follows the decrease in the thickness of the oil film until, at C, the oil film is no longer thick enough to provide adequate fluid separation of the surfaces, and the first significant incidence of asperity interaction occurs.

It will become important later in the report to understand what is implied by asperity interaction. Asperities on bearing surfaces are often covered by films of various types, and some theories suggest that in certain cases these films are fluid. An asperity interaction implies that asperities on the two surfaces have come close enough together for the local stresses to be influenced by the adjacent films, so that they are no longer determined by the bulk viscosity of the oil.

These first significant asperity interactions cause an increase in the coefficient of friction which appears in the Stribeck curve as an upward deviation from the almost straight descending line of the full-hydrodynamic zone. The increase in friction is associated entirely with localized stress on the asperities, so that the point C is also the point at which wear is likely to begin.

As the viscosity decreases further, the incidence of asperity interactions increases, and the coefficient of friction continues to increase from C towards B. At B viscosity has no effect on the coefficient of friction.

The ideal design should, theoretically, aim for operation at the point C of minimum friction and zero wear. In a real bearing there will be variations in bearing load, bearing speed (especially during start-up and shutdown), and temperature (resulting in viscosity change). It will therefore be impossible to hold the conditions at point C. In order to avoid any risk of wear it is desirable to keep any variations in $\eta N/P$ to the right of point C, and the traditional solution has been to use a viscosity higher than would be required for operation at C.

One effect of a higher viscosity oil is to increase the average frictional energy dissipation and thus the oil temperature. The viscosity will then decrease, and a new operating condition will be established which is still to the right of C.

With the use of adequate anti-wear techniques, the actual wear rate immediately to the left of point C will be low. In recent times the objective of energy conservation has been given much higher priority than previously, and for many applications the objective is now to design for operation at the lowest possible friction. Especially in automotive engines, where the operating speed and load vary over wide ranges, it is now the practice to select the viscosity so that some operation is permitted to the left of C.

The use of these lower-viscosity energy-saving oils has helped to achieve lower internal friction in automotive engines and lower fuel consumption.

It is obvious from Fig. 12 that the viscosity required to ensure operation at the point of minimum friction depends on the specific load and speed of the bearing. It will also depend on the surface roughness of the bearing surfaces, which ideally means the surface roughness after proper running-in.

The roughness of a surface is usually described either by the centre-line average (cla) or the root mean square (rms) of the asperity heights. The probability of asperity interactions is a function of the combined surface roughness σ, defined by equation (7).

$$\sigma = \sqrt{(\sigma_1^2 + \sigma_2^2)} \qquad (7)$$

where σ_1 and σ_2 are the separate rms roughnesses of the two surfaces.

If the distribution of asperity heights on both surfaces is Gaussian, then the first significant asperity interactions will occur at a separation much greater than σ. The first interactions are generally considered to occur when the ratio of the mean film thickness, h, to σ, called the lambda ratio Λ, is approximately equal to 3. At this point the incidence of asperity interactions is still small, and the effects on friction, flash temperature, and surface damage are negligible.

As the fluid film thickness decreases towards σ, the rate of asperity interactions and actual asperity contacts rises, and the extent of friction increase and surface damage increases. When the Λ ratio reaches 1.5, the amount of surface damage and wear can be expected to be significant, and boundary lubrication becomes an important factor in the bearing performance.

In a well run-in bearing the asperity height distribution ceases to be Gaussian and becomes skewed. This in itself does not markedly alter the above relationships, but at the same time the value of σ will have decreased, so that the actual film thickness for satisfactory operation will also be lower.

From the above discussion it is obvious that no absolute minimum value for viscosity can be defined, because the minimum value in any one application depends strongly on the bearing load, speed, and surface finishes. One recent survey of a wide variety of lubricated components has indicated that a practical minimum is 1cP. This is approximately the viscosity of water at 20°C and of kerosene at 50°–70°C.

From the previous chapter it is clear that even in present-day machines, the actual oil temperature in a high-temperature bearing may be well over 200°C. At such temperatures a fairly heavy lubricating oil such as an SAE50 (ISO 68) will have a viscosity of only about 2cP. The margin for increasing the temperature of bearings is obviously quite small unless the oil viscosity can be increased.

The Stribeck curve as shown in Fig. 12 originally applied only to hydrodynamic plain journal bearings, but the general principle applies equally to any hydrodynamically lubricated contact, in which a hydrodynamic wedge action is present, such as pad and step bearings and simple slider bearings. Figure 13 shows a 'Stribeck curve' produced experimentally in a pin-on-disc machine in which the pin was initially radiused and then run in.

The applicability of the above discussion to rolling bearings and gears, which are elastohydrodynamically lubricated, is not really valid for a number of reasons. According to equation (6), the film thickness is proportional to the 0.7 power of the viscosity in the entry zone in a cylindrical roller bearing, while the actual viscosity in the film at the point of maximum specific load depends on the speed and is an exponential function of pressure.

3.3 UPPER LIMIT OF VISCOSITY

The viscosity of a liquid varies monotonically with temperature, so that by analogy with the Stribeck curve, graphs can be drawn showing the variation of coefficient of friction for a specific lubricating oil at constant bearing speed and load, as the oil film temperature in the bearing changes. Figure 14 shows two such curves for mineral oils, developed theoretically from Fig. 13. The points B, B′, C, C′, D, and D′ represent the same bearing conditions as the points B, C, and D in Fig. 12, namely the boundaries between boundary and mixed lubrication and those between mixed and full fluid film lubrication.

The curves in Fig. 14 demonstrate clearly one of the main constraints on the use of liquid lubricants over a wide temperature range. If the viscosity is high enough to ensure full fluid film separation at a high temperature T_H, then the coefficient of friction becomes excessive at a lower temperature T_L.

Just as there is no general minimum viscosity to ensure full fluid film

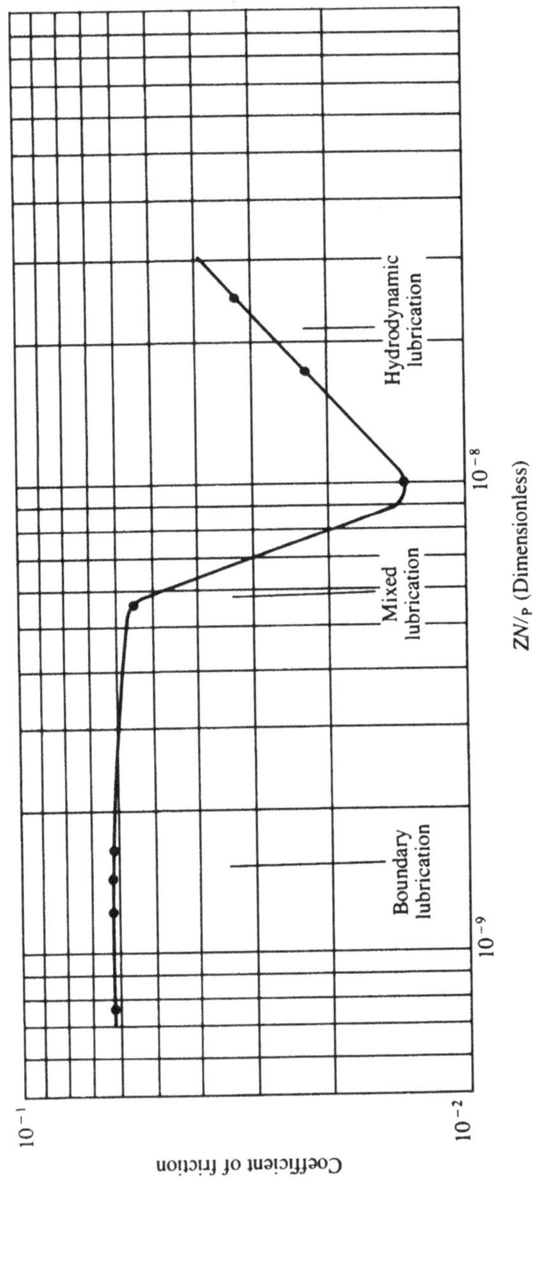

Fig. 13 'Stribeck curve' for pin-on-disc lubrication

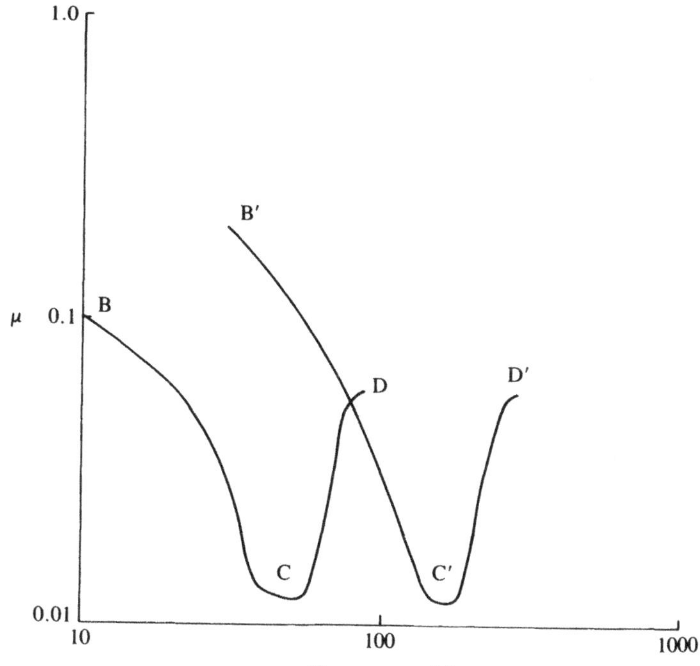

Fig. 14 Variation of friction with temperature

separation in journal bearings, there is also no general maximum viscosity to permit bearing operation at low temperature. The limitation on high viscosity depends only on the torque available to rotate the journal.

This problem has been familiar in aircraft engines since the earliest days of aviation. Aircraft piston engines are subject to severe constraints on size and weight. As a result bearing specific loads tend to be high, but temperatures are also high because space and weight are not available for large oil reservoirs or cooling systems. Aircraft piston engine oils are therefore more viscous than their automotive equivalents. For example, a standard 100 aircraft piston engine oil has a kinematic viscosity at 40°C of 260 cSt, compared with perhaps 60 cSt for an SAE 20 or 10 W/30 automotive oil.

This situation imposes constraints on the operation of aircraft piston engines. In cold weather starting may be difficult because insufficient torque is available, and techniques such as pre-warming of the oil or the whole engine, or dilution of the oil with fuel may be necessary. After start-up a lengthy warming-up period is usually necessary before the oil viscosity is low enough for the engine to provide full power for take-off. In a prolonged glide the oil temperature may drop so that the viscosity is too high for the engine to respond when power is required, and the engine must be frequently operated during the descent.

This indicates the nature of problems which may arise in high-temperature applications by virtue of the viscosity alone, but also gives some indications of possible techniques to deal with the problem. Nevertheless, any such techniques can only be palliatives, and the oil supplier will be squeezed by the need for high viscosity at high temperature and low viscosity at low temperature. It is important that oils for use at high temperature should have the least possible change in viscosity with temperature.

In Fig. 11 it can be seen that the rate of change of viscosity with temperature varies considerably between different oils. It is unfortunate that dimethyl silicones, which have the lowest change of viscosity with temperature of all known fluids, have serious disadvantages as lubricants. Of the other fluids, the plain mineral oil shows considerable decrease in viscosity with increasing temperature, and in general most fluids other than the dimethyl silicones become very much less viscous as the temperature rises.

3.4 VISCOSITY-INDEX IMPROVERS

In some respects the problem can be improved by the use of viscosity-index improvers. These are 'straight-chain' polymers such as acrylics and less commonly polyisobutenes. They have the property of increasing the viscosity of an oil much more at high temperature than at low temperature. The mechanism of this action is believed to be that at low temperature all the most effective viscosity-index improvers have limited solubility in the base oil, and tend to form tight spirals which have little effect on viscosity. At high temperatures they are more readily soluble in the base oil and open up into long spiral chains which approach their theoretical linear chain geometries and have a much greater effect on viscosity.

Earlier viscosity-index improvers such as those used in the manufacture of multi-grade automotive engine oils had a disadvantage known as temporary shear loss which resulted in ineffectiveness at shear rates higher than about $10^4\,s^{-1}$. Such shear rates are commonly found in high-speed bearings and gears, so that viscosity-index improved oils tended to lose the advantage of the additives when used in such components.

More modern viscosity-index improvers have been developed which are shear-stable to shear rates up to $10^6\,s^{-1}$, but it is not clear how completely these shear-stable additives have supplanted the earlier types. In one respect the problem is offset by the obvious fact that in bearings and gears with high shear rates there is a compensating tendency for the high operating speeds to produce higher oil film thicknesses.

A different problem with viscosity-index improvers is that the polymer chains are mechanically sheared at high shear strains, causing permanent shear loss. This problem and that of temporary shear loss have both been found to be more serious at higher temperatures. They can be helped, but not prevented, by the use of polymers with shorter chain lengths, but these must be used at higher concentrations.

3.5 PRESSURE EXPONENT OF VISCOSITY

In rolling-contact bearings and gears the primary lubricating mechanism is elastohydrodynamic lubrication, in which both the viscosity and the pressure exponent of viscosity of the oil are important. In an elastohydrodynamically lubricated contact very high pressure in the oil film causes an increase in lubricant viscosity in accordance with equation (8).

$$\eta_p = \eta_0\, e^{\alpha p} \qquad (8)$$

where α is called the pressure coefficient or pressure exponent of viscosity. The result in terms of the oil film thickness can be seen in equation (6).

The pressure exponent of viscosity, α, decreases with increasing temperature. The availability of data at high temperatures is limited, but figures for one mineral oil have been estimated as $3.0 \times 10^{-8}\,Pa^{-1}$ at 15°C and $1.7 \times 10^{-8}\,Pa^{-1}$ at 100°C. This represents a reduction of 29 per cent in film thickness for 85°C rise in temperature, according to equation (6), in addition to the very considerable reduction due to the lower ambient-pressure viscosity in the entry zone to the bearing or gear.

CHAPTER FOUR

Thermal Degradation

4.1 GENERAL

For clarity in discussing the high temperature behaviour of lubricants or any other materials, it is very important to keep to the strict definition of thermal degradation. This is deterioration which takes place by virtue of heat alone, with no reaction with any other substance, and particularly with no oxygen present. Any confusion about this definition leads to confusion in understanding high temperature effects.

Another expression which is commonly used in this context is thermal stability, which is resistance to thermal degradation in the absence of oxygen or any other reactive substance.

Oxidation always takes place at lower temperatures than thermal decomposition, although as will be seen later, the difference with fluorinated lubricants or phosphate esters may not be great. A knowledge of the thermal stability of a lubricant is valuable because it indicates the highest temperature at which it can be used, even in inert environments.

The rate of thermal decomposition theoretically increases as the temperature rises in accordance with the Arrhenius rate equation

$$k = Ae^{-E/RT} \tag{9}$$

where k is the reaction rate, A a constant, E the activation energy, R the gas constant and T the absolute temperature.

One feature of this relationship is that there is no temperature below which no thermal decomposition takes place. In other words, there is no temperature which can be defined as the absolute thermal decomposition temperature. There will, of course, be some temperature below which decomposition is undetectable, but this will depend on the criterion used to detect decomposition, the sensitivity of the analysis technique, and the elapsed time over which change is measured.

In spite of this, statements are frequently found in the literature such as 'esters of primary alcohols are stable up to 500–600°F'. Obviously such statements must be considered relative rather than absolute.

Nevertheless, there are indications that in oxygen-free conditions some liquids show no breakdown at all over very long periods, even when warm. It seems possible that at very low rates of initiation and the low molecular mobility at lower temperatures there may be a higher degree of radical recombination which effectively prevents chain reactions from occurring.

In thermal degradation the general tendency is for cleavage of the molecules to take place, giving smaller fragments. The process is similar to the 'cracking' of petroleum to give lower-molecular-weight products. The smaller fragments have higher volatility, and this has led to an empirical method of defining the thermal decomposition temperature, T_D. This is defined as the temperature at which the vapour pressure rises at the rate of 1.85 Pa/s.

This definition is the basis of the ASTM D2879 test for initial decomposition temperature of liquids by isoteniscope. It follows from this definition that the decomposition temperatures for a variety of organic compounds are those temperatures at which they all have identical rates of pressure rise.

The validity of this definition for a single chemical type of liquid such as mineral oils may be satisfactory, because they will tend to have the same general types of decomposition products. The validity when extended to liquids which are very different chemically is very doubtful, and it is probably significant that the ASTM test was originally intended primarily for petroleum products.

A valid definition for all classes of lubricant would need to show the temperature at which the rate of decomposition leads in an unacceptably short term to unsatisfactory lubrication performance. Unfortunately, no such definition has yet been found possible, and the user of a particular liquid at high temperatures may need to carry out his own testing to establish the time and temperature limits.

The mechanism of thermal decomposition is usually considered as beginning with the splitting off of an atom or group from a single molecule without any attack by a second molecule. The process is described in detail for mineral oils (hydrocarbons) in the next section, and subsequently for some of the synthetic oils.

No inhibitors have yet been found to be effective in improving thermal stability.

4.2 THERMAL DEGRADATION OF MINERAL OILS

Like oxidation, thermal decomposition of mineral oils proceeds by a free-radical chain reaction. Illogically, it is assumed that there would be no alkyl hydroperoxides present to initiate the free radical formation, although the same sample of lubricant operating in air would be assumed to contain them. Free alkyl radicals are known to be produced by the action of thermal energy or mechanical shear, so that in a high temperature bearing both sources will be available.

The formation of free alkyl radicals can follow a mechanism such as the following:

$$R{-}CH_2{-}CH_3 \rightarrow \dot{R} + \dot{C}H_2{-}CH_3 \qquad (10)$$

$\dot{R}$ and $\dot{C}H_2{-}CH_3$ are both alkyl radicals which are very reactive and will react with other hydrocarbon molecules by the removal of hydrogen atoms, thus creating new alkyl radicals.

$$R{-}CH_2{-}CH_3 + \dot{R} \rightarrow RH + R{-}\dot{C}H{-}CH_3 \qquad (11)$$

or

$$R{-}CH_2{-}CH_3 + \dot{R} \rightarrow RH + R{-}CH_2{-}\dot{C}H_2 \qquad (12)$$

The net result of equations (10), (11), and (12) is the creation of one stable molecule of reduced size, and two alkyl radicals, and can be summarized as follows

$$2\,R{-}CH_2{-}CH_3 \text{----} RH + \dot{C}H_2{-}CH_3 + R{-}\dot{C}H{-}CH_3 \qquad (13)$$

The attacks of free radicals on hydrocarbon molecules will continue until the radicals are converted into stable molecules by reaction or combination, such as

$$\text{(combination)} \quad \dot{R} + \dot{R} \rightarrow R{-}R \qquad (14)$$

$$\text{(reaction)} \quad R{-}\dot{C}H{-}CH_3 + \dot{C}H_2{-}CH_3 \rightarrow R{-}CH_2{-}CH_3 + \underset{\text{ethylene}}{CH_2{=}CH_2} \qquad (15)$$

The combination of two radicals can create a molecule larger than the original hydrocarbon, but the general trend is to create smaller molecules, such as RH and $CH_2{=}CH_2$ in the above equations. There are several reasons for this trend. One is that the smallest molecules created, such as ethylene, will tend to vaporize under the hot conditions of thermal degradation, so that they are removed from the liquid and can no longer be attacked. A second reason is that larger molecules are more liable to attack, probably because their bonds are more likely to be strained. A third reason is that a chain reaction is likely to proceed through hundreds of stages before termination occurs, and each stage generates at least one fragment of reduced size. There is, therefore, scope for the production of many smaller fragments before the single termination which may or may not produce a larger molecule.

It seems to be agreed by most authorities that the thermal decomposition of mineral oils or other hydrocarbons is not catalysed by metals. This is surprising because metal oxides are known to catalyse the similar thermal cracking of petroleum, and the surfaces of practical metals in contact with hot mineral oil lubricants will invariably be oxidized. In addition, the fresh dislocations and freshly-exposed metal of bearing surfaces have highly energized states which are known to cause dramatic degradation of hydrocarbon molecules. A possible partial explanation of this discrepancy is that the various laboratory investigations have used stationary metal coupons with stable oxide films.

4.3 THERMAL DEGRADATION OF SYNTHETIC OILS

4.3.1 Carboxylic esters

The carboxylic ester lubricants are synthetic chemical compounds formed by a condensation reaction between an alcohol and a carboxylic acid.

$$\underset{\text{alcohol}}{R{-}CH_2OH} + \underset{\text{acid}}{HOOC{-}R^1} \longrightarrow \underset{\text{ester}}{R{-}CH_2{-}CH_2{-}O{-}\overset{\overset{\displaystyle O}{\|}}{C}{-}R^1} + H_2O \qquad (16)$$

They are the most widely-used synthetic lubricants, having been used as gas turbine lubricants since the 1940s. Examples are di-2-ethylhexyl sebacate and various adipates and azelates. Their major advantages over mineral oils are lower low-temperature viscosity and greater thermal stability. This latter property has naturally led to considerable interest in the mechanism of their thermal decomposition.

The early ester lubricants have a long relatively straight alcohol structure, as shown in equation (16). On thermal degradation they produce a free acid and an olefine.

$$R{-}CH_2{-}CH_2{-}O{-}\overset{\overset{\large O}{\|}}{C}{-}R^1 \longrightarrow R{-}CH{=}CH_2 + H{-}O{-}\overset{\overset{\large O}{\|}}{C}{-}R^1 \qquad (17)$$

The mechanism for the breakdown has been shown to depend on an intermediate with a six-membered ring. Six-membered ring structures incorporating carbon atoms are more stable than rings with higher or lower numbers of atoms because the included angle of the hexagon, 120 degrees, is very close to the angle between unstrained bonds of the carbon atom. The stability of such a six-membered ring is also increased if 'resonance' can occur between different electronic configurations of the molecule, as is the case with the esters

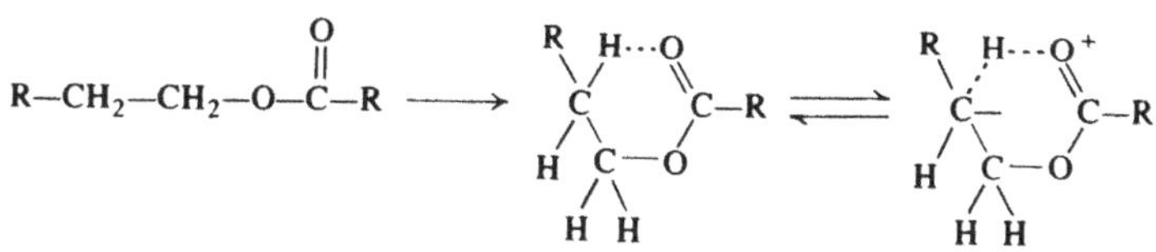

(A) (B)

(18)

The dotted lines in the ring formulae represent bonding tendencies rather than fixed bonds, and because they represent partial unbalancing of the bonding electrons they lead to a partial charge on the relevant atoms. The dotted line in structure (A) is due to a phenomenon called hydrogen bonding. The two structures (A) and (B) provide a medium for stabilization by resonance.

With thermal degradation, structure (B) splits to give a free acid and an olefine

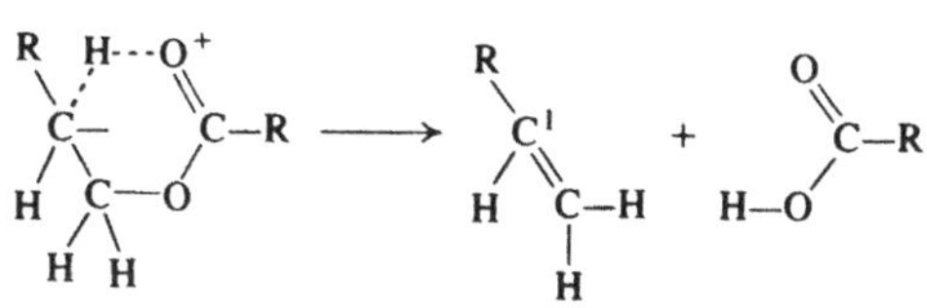

olefine acid

(19)

The carbon labelled 1 is the second from the hydroxyl group in the original alcohol (see equation (42)), and is called the beta position. It can be seen from equations (18) and (19) that hydrogen bonding to the oxygen of the C═O group can only give a six-membered ring if there is a hydrogen atom in the beta position. Substitution of the beta hydrogens by alkyl groups will prevent the above thermal decomposition mechanism from taking place.

Examples of alcohols which have no beta-hydrogen, and, therefore, give β-hindered esters, are trimethylolpropane, neopentyl glycol, trimethylolethane, and pentaerythritol. The last three are collectively known as neopentyl polyols because they are based on the hydrocarbon neopentane.

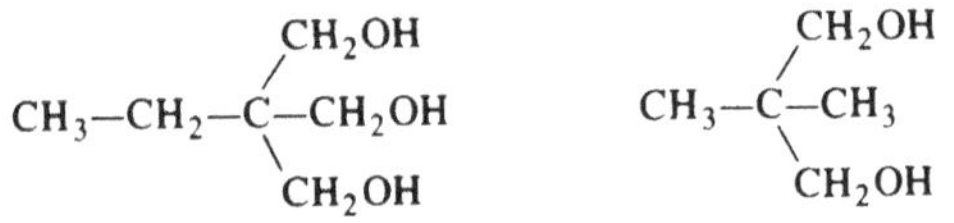

trimethylolpropane neopentyl glycol

trimethylolethane pentaerythritol

Esters of these alcohols with various carboxylic acids such as *n*-heptanoic acid are called 'hindered' esters, and have higher thermal stability than the simple di-esters.

Since the thermal breakdown through the six-membered ring intermediate is prevented in the hindered esters, thermal decomposition must take place by a free-radical mechanism, which requires greater activation energy and therefore higher temperatures. The formation of an olefine also requires a structural rearrangement, and a typical mechanism is shown in equation (20).

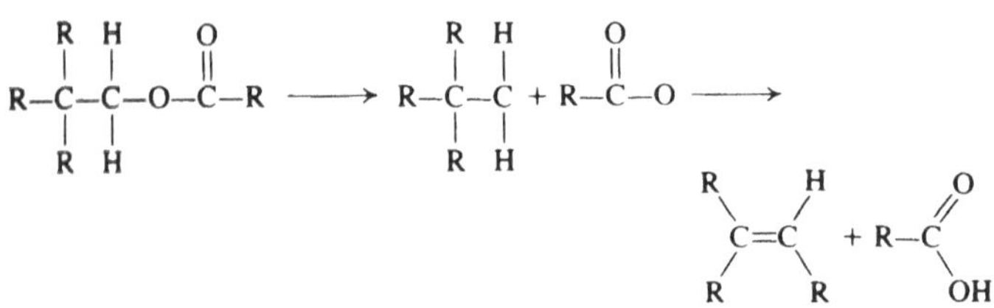

(20)

4.3.2 Polyglycols

Glycols are more-or-less straight-chain alkyl organic compounds with two OH groups. The simplest compound of the class is ethylene glycol

$HO—CH_2—CH_2—OH$, well known as an anti-freeze for cooling systems.

Polyglycols are linear polymers formed by a condensation reaction (i.e., removal of water) between glycols.

$$HO—CH_2—CH_2—OH + HO—CH_2—CH_2—OH \rightarrow \underset{\text{diethylene glycol}}{HO—CH_2—CH_2—O—CH_2—CH_2—OH} \quad (21)$$

The two glycol monomer residues are joined in the dimer by an ether linkage, so that polyglycols are in fact straight-chain alkyl polyethers. The simplest polymer is polyethylene glycol:

$$HO—CH_2—CH_2—O—[CH_2—CH_2—O]_n—CH_2—CH_2—OH$$

A variety of polymers can be produced and these are discussed in more detail in Chapter Eight. The most important factor for thermal stability seems to be whether the terminal OH groups are left unchanged or converted to ether linkages $R—O—CH_2—$. The diols, which terminate in OH groups, have nominal decomposition temperatures of about 230°C compared with about 275°C for the ether-terminated types.

The mechanism of thermal decomposition is believed to be similar to that of hydrocarbons, but the presence of a significant amount of combined oxygen probably enables some alkyl peroxide to be formed, permitting a mechanism like that of mineral oil oxidation.

The decomposition products are normally volatile and soluble in the oil, so that the polyglycols decompose cleanly, without deposit formation except at very high surface temperatures.

4.3.3 Phosphate esters

Phosphate esters are compounds which can be considered as condensation products of alcohols or phenols with phosphoric acid

$$3R—OH + (HO)_3P{=}O \rightarrow (RO)_3P{=}O + 3H_2O \quad (22)$$

The esters formed from alcohols, which are alkyl, are called trialkyl phosphates, while those formed from phenols, which are aromatic, are called triaryl phosphates. Esters are also used in which both alkyl and aryl groups are present and these are alkyl diaryl phosphates or dialkyl aryl phosphates.

The presence of an alkyl group results in lower thermal decomposition temperatures because the esters readily decompose to give an olefine and an 'acid ester', as in equation (23).

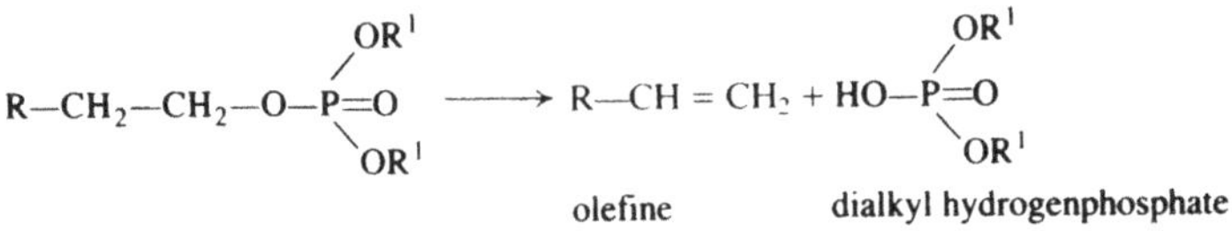

(23)

The presence of metals seriously reduces the thermal decomposition temperatures of phosphate esters, and in practice the triaryl phosphates can only be used to about 150°C and the alkyl diaryl or trialkyl phosphates to about 105°C.

4.3.4 Silicones

Polymeric compounds based on a chain structure of alternating silicon and oxygen atoms are called silicones. They also have organic side-chains, so the general structure is

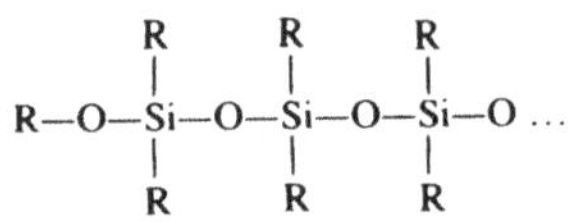

The side-chain R is a methyl group in the dimethyl silicones, but some of the methyl groups may be substituted in other silicones by longer alkyl chains or more commonly by phenyl, chlorophenyl, or trifluoropropyl.

The thermal decomposition of silicones begins with rupture of the silicon–oxygen bonds, but the free radicals formed then recombine to give either ring silicones or rubbery cross-linked compounds of higher molecular weight.

Theoretically the dimethyl silicones should be thermally stable to almost 540°C, but a practical limit is closer to 315°C. The practical limit for phenyl methyl silicones is a little higher, at about 370°C, provided that the end-groups of the chain are also phenyl groups. The chlorophenyl and trifluoropropyl silicones have thermal decomposition temperatures similar to those of the high-phenyl silicones, but they produce corrosive acidic decomposition products.

4.3.5 Silicate esters

The silicate esters are the theoretical reaction products of alcohols or phenols with silicic acid $Si(OH)_4$, and they have the general structure

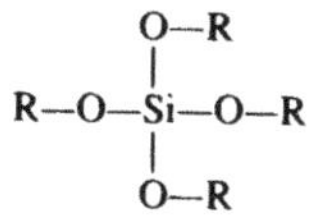

As with the phosphate esters, the organic groups may be alkyl or aromatic, giving tetraalkyl, tetraaryl, or mixed alkylaryl orthosilicates. There is also a sort of dimer silicate of the general structure

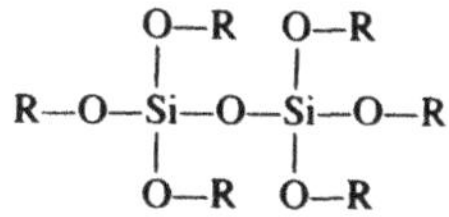

and these are properly called disiloxanes.

The alkyl silicates and disiloxanes decompose thermally in the same way as carboxylic esters to give an olefine and the free acid, but the free silicic acid is unstable and further decomposes to silica and water. The water released can then hydrolyse a molecule of ester to the original alcohol and silica.

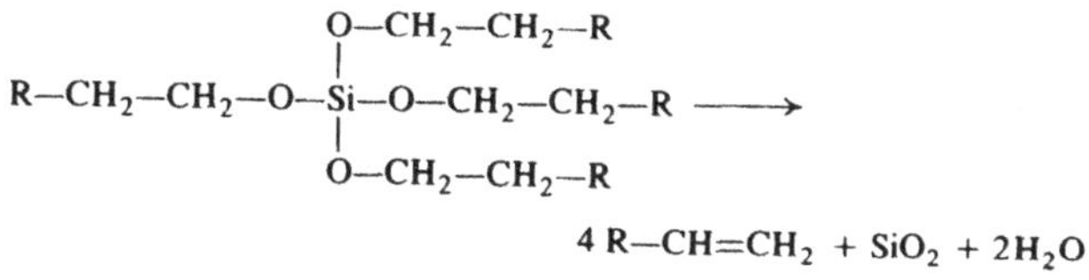

(24)

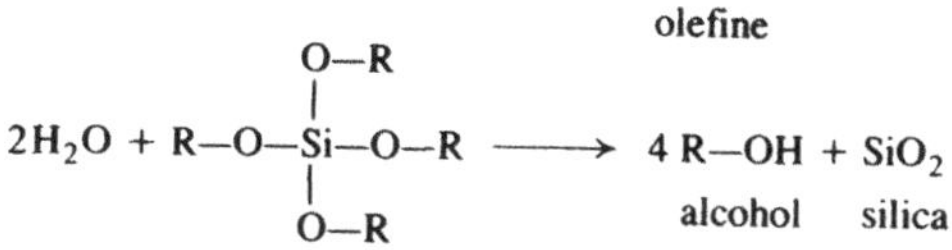

(25)

The aryl and alkylaryl silicates have a different decomposition mechanism, leading to a mixture of phenolic compounds and silica.

Typical thermal decomposition temperatures are 345°C for the alkyl esters and 450°C for the aryl esters.

4.3.6 Polyphenyl ethers

It was mentioned earlier that resonance between alternative structures of an organic compound has a stabilizing influence, and the supreme example of this is the six-membered benzene ring. Ether linkages are also thermally stable, so that the polyphenyl ethers, which consist of a number of benzene rings joined by ether links, are unusually stable.

Polyphenyl ethers will be described in more detail in Chapter Eight, but a typical example used for lubrication is the 5 phenyl 4 ether (5P4E) type.

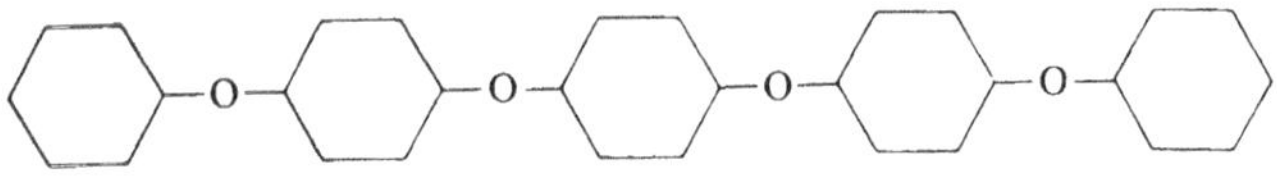

Typically, polyphenyl ethers have nominal initial decomposition temperatures in the range 440–485°C. The mechanism of thermal decomposition is apparently an initial rupture of an ether link to give a substituted phenyl and a substituted phenoxy radical

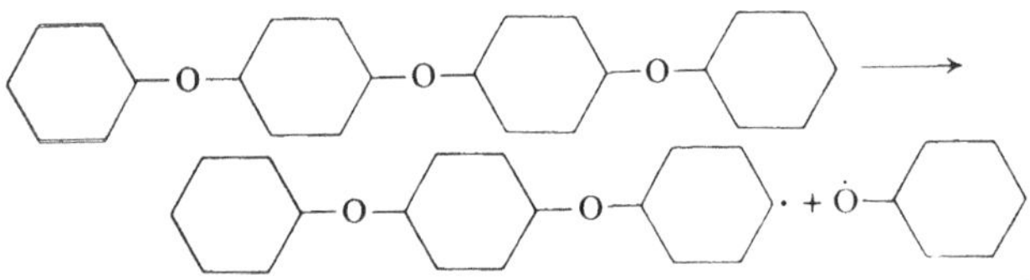

(26)

These radicals are very reactive, and further reactions follow which are in two categories. Removal of a hydrogen atom from a benzene ring can result in destabilization and complete disruption of the ring to produce small fragments such as carbon monoxide, carbon dioxide, water and aliphatic —C—C— structures. The hydrogen removed will stabilize the radicals, giving benzene, phenols or short-chain polyphenyl ethers

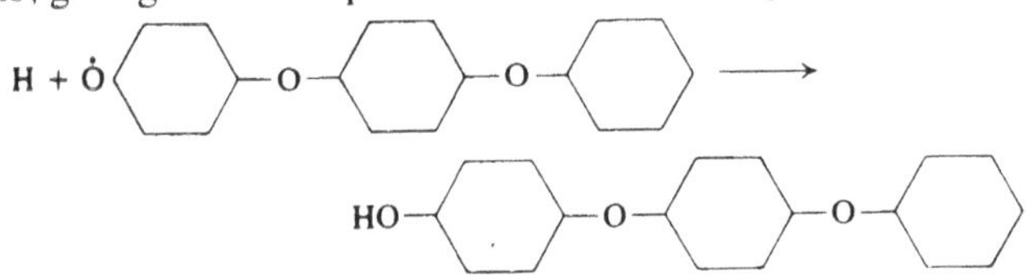

(27)

phenoxy-phenoxy-phenol

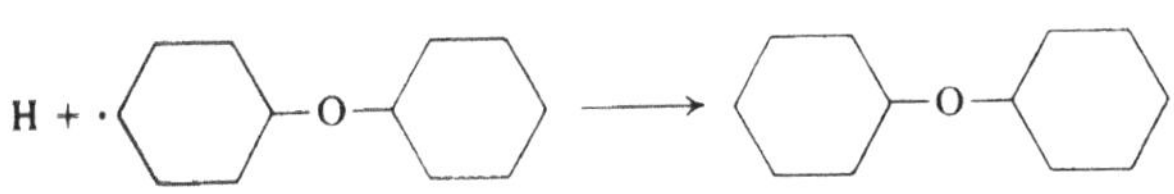

(28)

diphenylether

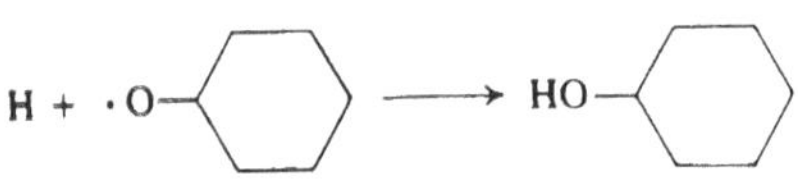

(29)

phenol

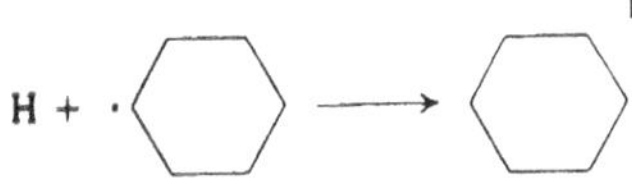

(30)

benzene

(It should perhaps be explained that if no atom or group is shown attached to one of the angles of a benzene ring, this indicates that the carbon at that point has a hydrogen atom attached to it but if a dot is shown it represents a free bond with nothing attached to it.)

The other category of reaction includes those in which radical attack results in recombination of ether chains. There is strong evidence that the longer chains produced by recombination are the most stable, since in one test over 40 per cent of the products had molecular weight higher than the starting compound.

4.3.7 Perfluoropolyalkyl ethers (PFAE)

The complete replacement of all hydrogen atoms by fluorine atoms in polyalkyl ethers generally similar to the ether-terminated polyglycols gives a variety of perfluoropolyalkyl ethers (PFAE). These compounds have great chemical stability and resistance to both thermal degradation and oxidation.

The most stable PFAE fluids have nominal thermal decomposition temperatures in the range 370–380°C, but they are sensitive to the presence of small quantities of chlorine and other unstable or destabilizing substances, such as residual hydrogen.

There appear to be at least two different mechanisms of thermal degradation of the PFAE fluids. The straight-chain molecules based on perfluorinated polypropylene oxide apparently degrade by a chain cleavage in which the cleavage of one molecule produces two smaller but still long-chain liquid fragments. The branched-chain molecules based on perfluorinated poly-isopropyl ether decompose by what is called an 'unzipping' mechanism. This is a mechanism in which one perfluoroisopropyl group at a time is detached from the polymer chain, and the process continues progressively along the chain. The decomposition products in this second case are gaseous.

The degradation of the PFAE fluids is increased in the presence of certain metals, and the nominal thermal decomposition temperature is reduced. However, these compounds attack metals at high temperatures and tend to remove surface oxide films, so it is possible that under these conditions some oxidation of the fluids is taking place.

4.3.8 Other synthetic oils

Several other synthetic oils have been investigated or used for lubrication, and some of these will be discussed in later chapters, but there is not enough information available to describe their thermal degradation.

CHAPTER FIVE

Oxidation

5.1 GENERAL THEORY OF OXIDATION

The general definition of oxidation in chemical terms is based on whether a chemical entity loses electrons in a reaction. By this definition fluorine is a powerful oxidizing agent because it has a strong tendency to attract electrons.

For the purposes of a discussion on oxidation in the context of lubrication, it is more helpful to use the older and more restrictive interpretation. Before the general adoption of the electronic concept of chemical reaction, oxidation meant simply the addition or substitution of oxygen into a molecule. This definition is adequate for most purposes in lubrication, but the modern definition does help in understanding the high resistance of fluorinated materials to oxidation. In electrochemical terms they are already highly oxidized.

Even in the more restrictive definition of oxidation, many different substances can bring about oxidation, by the transfer of oxygen from their own molecules to some other substance. One simple and common example is water, either as liquid or vapour, which can react with metals to produce surface oxide. Carbon dioxide will also react with metal surfaces to produce oxides, and has been added in trace quantities to inert gases in nuclear reactors to ensure the replacement of surface oxide removed by wear from bearing surfaces.

When wear of bearing surfaces completely removes the surface oxide from a spot on the surface of steel, the resulting freshly-exposed metal is highly active and has a strong attraction for oxygen. Almost any oxygen-containing molecule may react with such a surface and re-oxidize it, including synthetic oils such as polyglycols and esters, lubricant additives, and even some compounds such as sodium nitrite, which are normally reducing agents.

The most common oxidizing agent, or oxidant, is atmospheric oxygen, which is almost universally present in lubrication, and most of the discussion in this chapter will be related only to atmospheric oxidation. Where oxygen is excluded from a lubricated system, severe wear problems can arise due to removal and non-replacement of the surface oxide from bearing metals, and special remedial actions will normally have to be taken to ensure the maintenance of a protective oxide film.

Like most chemical reactions, oxidation can be accelerated by catalysts, and the oxidation of lubricants is often strongly catalysed by bearing surfaces, especially when they have been freshly worn.

A great deal of research has been carried out on the oxidation of mineral oils because of the considerable economic importance of the problem. This research has in fact helped to establish some of the fundamental understanding of oxidation processes. The next section will therefore describe in detail the oxidation of hydrocarbons such as mineral oils.

5.2 OXIDATION OF MINERAL OILS

5.2.1 Low-temperature oxidation

The low-temperature oxidation of mineral oils takes place through a mechanism involving free radicals.

Free radicals are very reactive and have a limited life, although their instability varies, some free radicals being considerably less stable than others.

The hydrocarbon oxidation process takes place in three stages, initiation, propagation, and termination. In the initiation stage a hydrocarbon molecule loses an atom of hydrogen to give a hydrocarbon free radical. In the propagation stage a hydrocarbon free radical reacts with a molecule of oxygen to form a peroxy radical. The peroxy radical then reacts with a fresh hydrocarbon molecule to form a hydroperoxide molecule and a new hydrocarbon free radical. In the termination stage two free radicals interact to produce one or more stable molecules.

The three stages are represented as follows

$$\text{Initiation} \quad \underset{\text{hydrocarbon}}{RH} \xrightarrow{\text{activation}} \underset{\substack{\text{hydrocarbon}\\ \text{free radical}}}{R\cdot} + \underset{\substack{\text{hydrogen}\\ \text{atom}}}{H\cdot} \qquad (31)$$

$$\text{Propagation} \quad \underset{\text{free radical}}{R\cdot} + \underset{\text{oxygen}}{O_2} \rightarrow \underset{\text{peroxy radical}}{ROO\cdot} \qquad (32)$$

$$\underset{\text{peroxy radical}}{ROO\cdot} + \underset{\text{hydrocarbon}}{RH} \rightarrow \underset{\text{hydroperoxide}}{ROOH} + R\cdot \tag{33}$$

$$\text{Termination} \quad \underset{\text{peroxy radicals}}{2\,ROO\cdot} \rightarrow \text{various stable molecules} \tag{34}$$

Each of the stages will now be considered in more detail.

5.2.1.1 Initiation stage

There is a lack of clear understanding about the mechanism of the initiation stage. All authorities agree that oxidation chains are often initiated by traces of hydroperoxide, according to equations (35) and (36)

$$ROOH \rightarrow ROO\cdot + H \tag{35}$$

$$ROO\cdot + RH \rightarrow ROOH + R\cdot \tag{36}$$

This mechanism certainly takes place when alkyl hydroperoxides ROOH are present in a hydrocarbon. It is also certain that alkyl hydroperoxides are normally present in any sample of hydrocarbon such as a mineral oil. What is not fully established is whether the presence of hydroperoxides is necessary for the initiation of oxidation, or whether the hydroperoxides are themselves the product of a first reaction stage. Alkyl hydroperoxides are a side-product of the propagation stage, so that any other initiation mechanism will subsequently ensure an adequate supply of hydroperoxides.

This subject is of more than academic importance, since substances, such as phenothiazine, are readily available which destroy alkyl hydroperoxides. If alkyl hydroperoxides are the only initiators of oxidation, then it would be possible to prevent oxidation completely by the use of phenothiazine. This is not so; phenothiazine delays oxidation, but does not prevent it.

It seems much more likely that oxygen can react directly with a hydrocarbon. The probable mechanism is based on the fact that the oxygen molecule itself behaves like a diradical, because it has two unpaired electrons. As a result it will attack a hydrocarbon and remove hydrogen.

$$RH + \cdot O{-}O\cdot \rightarrow R\cdot + \underset{\text{hydroperoxy radical}}{HOO\cdot} \tag{37}$$

The alkyl radical can then react with oxygen to form an alkylperoxy radical

$$R\cdot + \cdot O{-}O\cdot \rightarrow R{-}O{-}O\cdot \qquad (38)$$

while the hydroperoxy radical can also attack a hydrocarbon molecule to produce an alkyl hydroperoxide.

$$RH + HOO\cdot \rightarrow ROOH + H\cdot \qquad (39)$$

Initiation of oxidation is strongly affected by the types of hydrocarbon or other compound present in a mineral oil. Mineral oils produced by very mild refining processes contain both oxidation initiators and oxidation inhibitors. More severe refining tends to remove initiators more than inhibitors, and at an intermediate stage of refining severity, optimum oxidation resistance is obtained.

As the severity of refining increases from that point the oxidation resistance decreases until with the most severe refining processes the oxidation resistance is quite poor. This effect is shown in Fig. 15. Fortunately at the same time the effectiveness of anti-oxidants improves, so that the modern tendency is to employ severe refining processes and to use anti-oxidants to ensure adequate oxidation resistance of the products.

The initiation stage of low-temperature oxidation is encouraged by higher temperatures.

5.2.1.2 Propagation stage

The simplest propagation mechanism is shown in equations (32) and (33). The reaction in equation (33) produces an alkyl hydroperoxide ROOH and an alkyl radical R·. The alkyl hydroperoxide is available to initiate a fresh oxidation sequence. The alkyl radicals are highly reactive and react rapidly with fresh oxygen in accordance with equation (32). The process is thus self-propagating and a chain reaction occurs, one molecule of oxygen being absorbed and one molecule of hydrocarbon being attacked in alternate steps.

In a typical autoxidation process the chain reaction may occur over a hundred times before termination. As a result, the overall reaction rate is dominated by the propagation stage.

Apart from the reactions shown in equations (32) and (33), an alternative propagation mechanism can follow spontaneous decomposition of an alkyl hydroperoxide

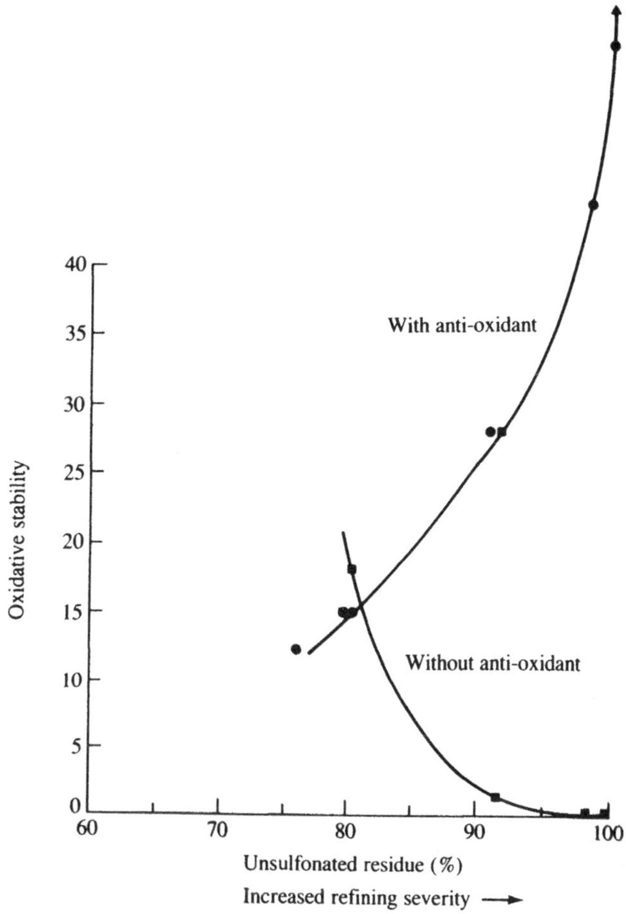

Fig. 15 Effect of refining severity on oxidation

$$\underset{}{ROOH} \rightarrow \underset{\text{alkoxy radical}}{RO\cdot} + \underset{\text{hydroxy radical}}{\cdot OH} \qquad (40)$$

$$RO\cdot + RH \rightarrow R\cdot + \underset{\text{alcohol}}{ROH} \qquad (41)$$

Since ROOH is produced by the reaction of equation (33), this represents a branching of the chain and a doubling of the chain propagation.

5.2.1.3 Termination stage

The basic termination stage is shown in equation (34), but many alternative terminations are possible in which any pair of free radicals combine together to form a larger stable molecule. Examples are

$$\underset{\text{}}{R\cdot + R\cdot \rightarrow} \underset{\text{hydrocarbon (dimer)}}{R{-}R} \tag{42}$$

$$R\cdot + RO\cdot \rightarrow \underset{\text{ether}}{ROR} \tag{43}$$

$$R\cdot + ROO\cdot \rightarrow \underset{\text{alkyl peroxide}}{ROOR} \tag{44}$$

Chain termination can also take place by reaction of a free radical with an unsaturated molecule (olefine) or aromatic molecule to produce a large free radical.

$$\underset{\text{aromatic hydrocarbon}}{R\cdot + PhH} \rightarrow \underset{\text{aromatic radical}}{PhR\cdot + H\cdot} \tag{45}$$

Certain non-hydrocarbons, such as hindered phenols or amines, are readily attacked by free alkyl radicals to form stable compounds.

5.2.2 Hot oxidation of hydrocarbons

The effect of heat on all chemical reactions is to increase reaction rates. In the case of mineral oil oxidation, there are also two processes which are believed by some authorities to take place at high temperatures but not at low temperatures. The first of these is the branching mechanism shown in equations (40) and (41), which has been described as taking place mainly above 100°C.

The second mechanism is an alternative to the peroxide mechanism described previously. The basic reaction is a direct oxidation of hydrocarbon molecules to give free radicals and water.

$$4\,RH + O_2 \rightarrow 4R\cdot + 2\,H_2O \tag{46}$$

The free radicals produced can take part in all the types of reaction described previously, but because of their higher concentration than in the peroxide process there is a greater probability of them reacting together to produce polymers, as in equation (42).

If these theories are correct, there are three factors which cause higher oxidation rates at high temperatures:

(1) The general increase in reaction rates due to high temperatures.
(2) The chain branching mechanism.
(3) The direct attack of oxygen on hydrocarbon molecules.

One effect of the higher oxidation rates at high temperature may be that the actual rate of deterioration of an oil is controlled by the availability of oxygen and not by the reactivity of the oil. This is particularly true in the confined spaces of bearings.

The problem of oxygen availability is an important factor in oxidation tests. In order to obtain useful results in a short time, oxidation tests are always carried out at raised temperature. In a static test, because oxygen is a gas with limited solubility in the liquid oil, the supply of oxygen would be limited by the diffusion and solution rates of the oxygen in the oil. It is therefore necessary in hot tests to pass the oxygen or air at a high rate through the heated oil to ensure that sufficient oxygen is present.

The two phases, gaseous oxygen and liquid oil, meet at the surfaces of the gas bubbles. The supply of oxygen by diffusion is controlled by solution at the bubble surfaces, but oxidation will also take place to some considerable extent at the gas–liquid interface. As a result the nature of the mixing must be carefully controlled. It also becomes difficult to correlate the results of different test methods, because of the dominant influence of the nature of the mixing process used.

5.2.3 Effects of catalysts

The effects of metal catalysts on oxidation have been clearly established. Most research investigations have used metals such as copper, cobalt and manganese dissolved in the oil in the form of metal–organic compounds and chelates as catalysts.

A much more important problem for high-temperature lubrication is catalysis by bearing surfaces. This has been shown to occur especially when the bearing surfaces are being worn in the presence of the oil. Under these circumstances the bearing surfaces are highly activated by the presence of freshly-exposed metal and dislocations.

In principle, catalysts increase the rate of reactions without being themselves changed. In the case of catalysis by bearing surfaces, two factors arise which affect the nature of the products as well as sometimes altering the surfaces. The first of these is the fact that the oxidation reactions take place preferentially adjacent to the bearing surfaces. The second is that the oxidation is confined to a thin film, which is, in any case, at a higher temperature than the bulk. The presence of oxidation processes adjacent to the surfaces can also affect the state of oxidation of the surfaces, especially with silicones and perfluoroalkyl ethers.

The result of both these factors must be that there is a higher probability of the oxidation reactions involving interactions between two radicals or attack on the products of previous oxidations. The overall extent of deterioration may not be any greater than is predicted on energy grounds, but the degradation products are likely to consist of small quantities of more highly-reacted substances.

5.2.4 Anti-oxidants

In theory there are three ways in which the extent of oxidation of the oil can be reduced. The first is by the use of reducing agents which preferentially react with the oxygen available and remove it from the oil. This could only be practical where the amount of oxygen present is already limited either by geometry or by operation in an inerted environment. In any other situation the amount of oxygen available is so great that an excessive amount of the anti-oxidant would be needed.

The second way is to use a material which destroys the initiating compound. The action of phenothiazone in destroying alkyl hydroperoxides is an example of this process. It would presumably be effective in preventing the normal low-temperature process based on equation (35), but not on the suggested hot process based on equation (46).

The third method is to use a substance which will interfere with the oxidative chain reaction by reacting rapidly and preferentially with one of the intermediate compounds taking part in the chain. The simplest type of action takes the place of equation (33), as follows

$$\text{ROO}\cdot + \text{AH} \rightarrow \text{ROOH} + \text{A}\cdot \qquad (47)$$

$$\text{ROO}\cdot + \text{A}\cdot \rightarrow \text{stable products} \qquad (48)$$

where AH is the antioxidant. An example of this type of antioxidant is 2,6-di-tertiary-butyl-4-methylphenol, which gives the following reactions.

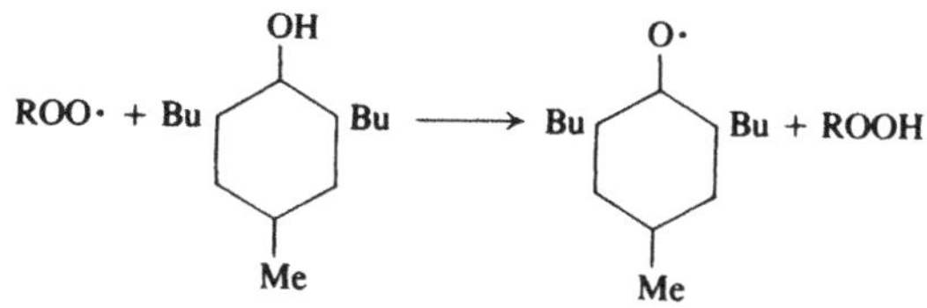

(49)

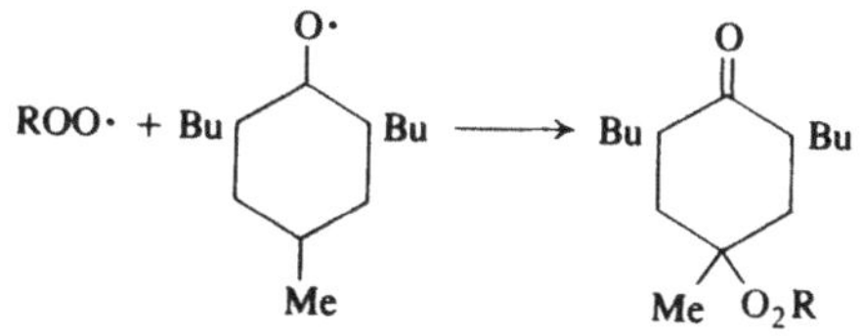

(50)

In addition to the substituted phenols, various amines can react with the alkylperoxy radicals. One of the best types is that of the para-phenylenediamines, which will react in rapid succession with two alkyl-peroxy radicals.

$$2\,ROO\cdot + RNH.C_6H_4\,NHR \rightarrow 2\,ROOH + RN{=}C_6H_4{=}NR \qquad (51)$$

The value of both these types of chain-breaker antioxidant is limited for two reasons. The first is that they generate alkyl hydroperoxides, which are themselves chain initiators. The second is that they can only interrupt a chain which has already started, so that they are not very useful where there is a high rate of chain initiation.

Apart from phenothiazine, several other substances, including various organic sulphur compounds, destroy hydroperoxides by reactions which do not produce free radicals. The most important of these are the zinc dialkyl dithiophosphates and similar compounds, which are very widely used not only as antioxidants but also as corrosion inhibitors and extreme-pressure additives.

Mixtures of different types of antioxidant are often more effective than the sum of their separate efficiencies. This synergistic effect is particularly notable when a chain-breaker such as a substituted phenol is used in conjunction with a hydroperoxide destroyer.

Two other types of substance exist which inhibit oxidation. These are the ones which form a protective film on bearing surfaces to prevent surface catalysis, such as phosphorus compounds, and those which chelate or otherwise remove soluble heavy metal compounds which can act as oxidation catalysts.

All types of anti-oxidant are destroyed in use, so that the pattern of oxidation consists of an induction period during which little or no

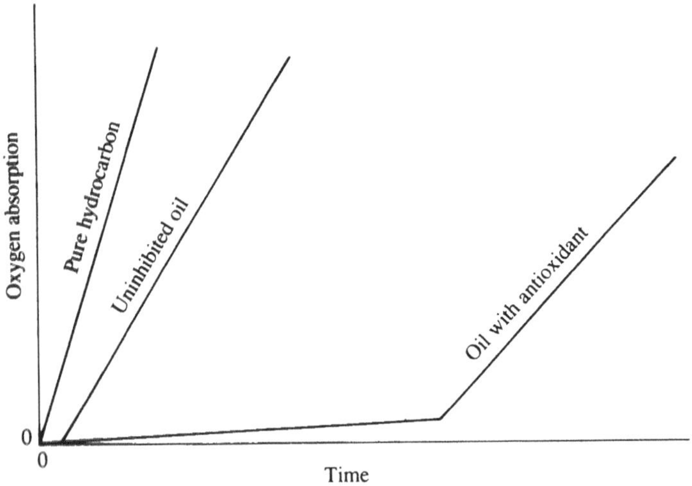

Fig. 16 Typical oxidation curves

oxidation of the oil takes place, followed by a period of rapid oxidation once the antioxidant is used up. This situation is shown in Fig. 16. A similar pattern exists with mildly-refined mineral oils without added antioxidants, supporting the view that such oils contain natural oxidation inhibitors.

Some of the products of mineral oil oxidation also act as inhibitors, so that the oxidation process is eventually self-inhibiting. This is not of practical importance in lubrication because the extent of degradation required for self-inhibition would be unacceptable for lubrication purposes.

Satisfactory antioxidants exist for most applications of mineral oils, but they become ineffective at higher temperatures, mainly due to their own reactivity, and none of them is suitable for the very high temperature synthetic oils.

5.2.5 Oxidation products

Most of the hydrocarbon oxidation takes place in the chain propagation stage of the oxidation process. The products of this stage are the alkyl hydroperoxides and free alkyl radicals, but the latter are too reactive to remain free and promptly react with further oxygen.

The only significant products according to this scheme are therefore the hydroperoxides. In practice, however, high concentrations of hydroperoxides are never found, so it is clear that other reactions must take place.

It has been shown that the alkyl peroxy radicals formed in equation (32) can combine to form tetroxides

$$2\,ROO\cdot \rightarrow ROOOOR \qquad (52)$$

This reaction is reversible, but at high temperatures the tetroxides are unstable and decompose to form ketones and alcohols

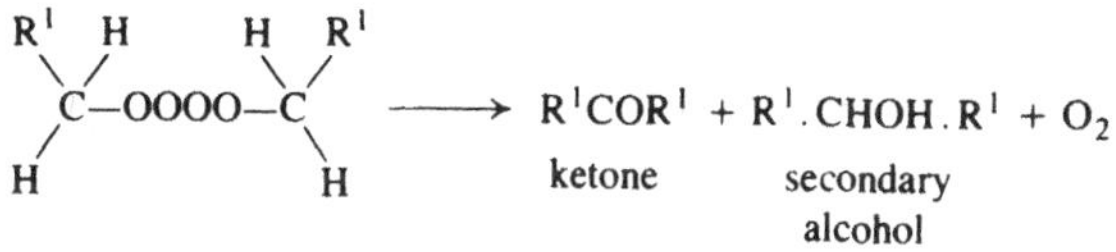

(53)

The secondary alcohols are themselves fairly readily oxidized to ketones. Ketones and alcohols are certainly found in mineral oil oxidation products, but more common products are aldehydes and acids. It is probable that the former arise from a free radical reaction with an alkoxy radical such as

$$RCH_2O\cdot + ROO\cdot \rightarrow \underset{\text{aldehyde}}{R.CHO} + ROOH \qquad (54)$$

and aldehydes are readily oxidized to carboxylic acids.

$$2\,R.CHO + O_2 \rightarrow \underset{\text{carboxylic acid}}{2\,R.COOH} \qquad (55)$$

Severe oxidation of mineral oils leads to an increase in viscosity and eventually to sludges, as well as acidity. The simple alcohols, aldehydes, ketones, and acids do not account for either thickening or sludge formation. The thickening is probably explained by the production of larger molecules by free radical combination. Equation (42) gave an example of a large molecule R—R formed by combination of two alkyl free radicals R·. Such addition reactions may be common in hot oxidation where free radical concentrations are high.

The extent of oxidation of an oil can be assessed by smell, increased acidity, increased carbonyl absorption in the infra-red spectrum, increased viscosity and direct testing for the presence of high-molecular-weight products and sludges.

The alcohols, ketones, and acids produced by mild oxidation contribute to what has been described as a 'fruity smell', whereas more severe oxidation tends to generate a more sour smell. Unfortunately these smells can be masked by variations in additives.

The acidity which develops from oxidation is a weak organic acidity which is usually measured by a Total Acid Number (TAN) test such as IP139 (ASTM D974) or IP177 (ASTM D664). The TAN of an unused oil is usually less than 2 mg.KOH/g., and this will rise fairly slowly as long as antioxidant is present. A rise to 10 mg.KOH/g. would indicate that fairly severe oxidation has taken place.

The acidity is due mainly to carboxylic acids formed by oxidation, unless the oil is from an engine using sulphur-containing fuel, in which case some sulphur acids may be present. The sulphur acids are strong acids and can be differentiated from the weak organic acids by the Strong Acid Number (SAN) methods in the above tests.

Carboxylic acids, ketones and aldehydes all contain the carbonyl group C═O, and this group causes strong absorption at about 5.8 μm in the infra-red spectrum. Unfortunately any acid or ester additives will mask the absorption due to oxidation, so a combination of acidity and spectroscopic tests is desirable.

The change in viscosity can be measured by the usual capillary viscometer tests of IP71 (ASTM D445). Direct measurement of the quantity of heavy molecules or sludge can be made by testing for Pentane-Insoluble Material, but this test will also detect any carbonaceous products from poor combustion if the oil is from an internal combustion engine.

5.3 OXIDATION OF SYNTHETIC BASE OILS

The published literature on oxidation studies of synthetic oils is mainly concerned with the various esters, because of their widespread use, and the very high-temperature oils such as polyphenyl ethers and perfluoropolyalkyl ethers. Even these studies have been far less detailed than those into mineral oil oxidation.

The evidence available suggests that polyglycols and the carboxylic ester lubricants, such as the gas turbine lubricants di-2-ethylhexyl sebacate and the more stable trimethylolpropane heptanoate probably oxidize by peroxide mechanisms similar to those of mineral oils. The results are also similar, leading to acidity, thickening, and sludge formation, but the viscosity increase seems to be due mainly to polymerization.

The polyphenyl ethers tend to be split by attack at the ether linkages rather than by rupture of their aromatic rings. This produces phenoxy or substituted phenoxy radicals which propagate the oxidation. The oxidation of polyphenyl ethers is actually inhibited by some metals and metal oxides.

With the perfluoropolyalkyl ethers oxidation appears to produce a breaking of the polymer chains, to generate lower molecular-weight and less viscous products. Even at temperatures as high as 344°C oxidation is very limited in the absence of metals, but several metals catalyse a chain cleavage mechanism. The products are generally not highly oxidized, and it seems possible that the main breakdown is a thermal degradation which is only partly affected by an oxygen atmosphere.

Oxidation of the dimethyl silicones takes place mainly at temperatures greater than 210°C in the absence of catalysts. The mechanism is not clear, but there are indications that methyl groups oxidize to formaldehyde, carbon dioxide and formic acid, while the silicon—oxygen chain first forms a highly crosslinked polymer. Eventually, after further oxidation the main reaction product is silica (silicon dioxide SiO_2). This would suggest that once a particular molecule is attacked it degrades rapidly so that no stable intermediate oxidation products are formed. As with most organic compounds, metals catalyse the oxidation, and bearing metals apparently cause oxidation even as low as 100°C.

The best antioxidants for the carboxylic esters are phenothiazine and certain derivatives. As described previously, this compound acts by destroying the hydroperoxide chain initiators. The rate of loss of phenothiazine is unexpectedly low, and it has been suggested that it forms an additive compound with oxygen which subsequently loses the oxygen to regenerate the phenothiazine.

The search for antioxidants for the higher-temperature lubricants has not been very successful. A few have been found which are specific to certain types of fluid, such as tetraphenyltin for the polyphenyl ethers and a perfluorinated substituted triphenyl phosphine for the perfluoropolyalkyl ethers, but it is not clear that any of these is yet adequate for service use.

5.4 COMBUSTION

Combustion is an extreme type of oxidation, in which the heat generated by the oxidation processes is sufficient to ensure continuing oxidation at a high rate.

At present combustion of lubricants in service is uncommon. It takes place in some hot metal-working processes such as continuous casting when an organic lubricant is used and comes into contact with very hot or molten metal. Some oil is also burned in internal combustion engines, either by entrainment with the fuel, or in the surface of oil films in the combustion chambers.

The matter of lubricant combustion only becomes of real significance if a fuel is used as a lubricant or if it is intended that a lubricant should be burned with the fuel, and this latter possibility is discussed later. At this stage it may be useful to mention some characteristics of combustion which need to be considered in such a case.

Oil vapours or mists will burn at concentrations between about 3 and 17 per cent of oil in air by weight, but for efficient combustion a very high oxygen availability is necessary. A typical hydrocarbon oil needs the same fuel/air ratio as a hydrocarbon fuel, namely more than 14.7:1 air/oil ratio on a weight-to-weight basis. If insufficient oxygen is present, there will be incomplete combustion, giving tarry or solid combustion products or even carbon. Because of the need for high oxygen availability, combustion normally takes place only in the vapour phase or at the surface of a solid or liquid.

In the latter case it may be difficult to maintain combustion if the heat produced by oxidation of the surface molecules is not enough to raise the temperature of the adjacent bulk solid or liquid high enough for it to ignite. Most liquid lubricants are also too high in molecular weight to be readily vaporized, so adequate mixing with air is obtained by producing a fine spray of the liquid.

The converse is that, even at temperatures high enough for combustion, an oil will not burn in the absence of large amounts of oxygen, so that an oil can lubricate satisfactorily even above its flash point or spontaneous ignition temperature provided that the amount of oxygen present is low enough.

In the oil film in a lubricated contact the availability of oxygen is generally very low, so that ignition of the lubricant is almost impossible, however high the flash temperature or even the average contact temperature may be. It is only when a hot spot moves out of the contact zone into a position where it is exposed to oxygen-rich atmospheres containing oil vapour or droplets above the spontaneous ignition temperature that ignition becomes a serious risk. Such conditions can occur in air compressors and will become more likely as system temperatures rise.

CHAPTER SIX

Effects of Heat on Boundary Lubrication

6.1 GENERAL

The boundary regime in lubricated sliding occurs when the load between lubricated contacts is carried by direct interactions between the surfaces, and not through a largely-uniform continuous film of lubricant.

The meaning of boundary lubrication is quite clear to all who are interested in lubrication, but it has proved very difficult to define it satisfactorily. The idea which must be conveyed is that the quality of lubrication is determined by the nature of films or layers at or near the load-bearing surfaces, and that those films or layers differ from the bulk of the lubricating fluid other than simply in temperature and pressure.

The concept is perhaps shown most clearly by means of the Stribeck curve in Fig. 12, where the region between A and B is the zone of boundary lubrication. The region between B and C is a zone of mixed lubrication where both boundary lubrication and hydrodynamic lubrication contribute to the overall performance.

The chemical and physical factors involved in boundary lubrication are very complex. Table 2 lists twenty-four phenomena which can influence boundary lubrication, and others could certainly be added. Although all of these factors can influence the general situation in the boundary regime, only about six of them are generally recognized as making an important contribution. These are adsorption, chemisorption, reaction of boundary additives, polymerization, micro-elastohydrodynamic lubrication, and micro-rheodynamic lubrication. These six factors will be described individually in relation to the effects of high temperature.

Before that it may be useful to give a description of the various types of chemical additive which are used to enhance boundary lubrication.

6.2 BOUNDARY LUBRICATION ADDITIVES

The additives used to improve boundary lubrication are generally classified under three headings. Depending on the effect they are intended to

Table 2 Phenomena affecting boundary lubrication

1	Enrichment of oxygen and other non-metallic 'contaminants' near the surface of a metal.
2	Concentration or dilution of specific metallic elements at the load-bearing surfaces (e.g., selective transfer).
3	Adsorbed films.
4	Effect of adsorbed films on yield stress (Rehbinder effect).
5	Possible viscosity increase in liquid adjacent to surfaces.
6	Elastic deformation of asperities and substrate.
7	Plastic deformation of asperities.
8	Residual stress.
9	Work hardening.
10	Fatigue.
11	Adhesion.
12	Melting.
13	Welding.
14	Chemisorption.
15	Decomposition of lubricating oil and additives.
16	Reaction of boundary additives.
17	Generation of free hydrogen followed by reduction of surface materials, especially carbides and oxides.
18	Generation of reactive organic radicals and reaction with surface materials.
19	Polymerization.
20	Flash temperatures.
21	Micro-elastohydrodynamic lubrication.
22	Micro-rheodynamic lubrication.
23	Electron emission.
24	Pressure gradients in the lubricant.

achieve, they are called friction-reducing (lubricity or oiliness) additives, anti-wear additives and extreme-pressure (EP) additives.

The terms 'lubricity additives' and 'oiliness additives' are two of the many ill-defined and ambiguous expressions which plague lubrication technology. Because lubrication properly means reduction or control of friction, 'lubricity additives' should mean friction-reducing or friction-controlling additives. This conflicts with the term 'friction modifiers', which is used specifically for additives which modify the transition between dynamic and static friction in the clutch plates of an automatic transmission.

The term 'oiliness additive' is sometimes used synonymously with 'lubricity additive' and sometimes contrasted with it. They are both sometimes called 'film-forming additives' to distinguish them from extreme-pressure additives, but in fact the latter also form films. Either 'oiliness' or 'lubricity' may be used to mean anything from friction-reducing to mild EP.

In this report it is proposed to use the expression 'friction-reducing' for additives whose primary purpose is the reduction of friction in boundary lubrication. These are generally fairly unreactive and only slightly polar compounds with long alkyl carbon chains in their molecules. Examples are natural oils and fats such as rapeseed oil, long-chain amides and alcohols such as stearamide and lauryl alcohol, and similar synthetic products.

The friction modifiers are much more specific in their action than the friction reducers. Their primary use is in lubricated clutches where a simple reduction in friction is not desirable. Their purpose is to ensure smooth engagement of the clutch plates, and particularly to prevent any stick-slip phenomena which would lead to judder, 'squeal', or 'squawk'. The compounds used include sulphurized fats and esters as well as some more specialized chemicals such as esters of dimerized fatty acids and N-acyl sarcosines.

Because the asperity interactions involve high local stresses, wear normally takes place in the boundary regime, and an objective of boundary lubrication is to reduce wear as well as friction.

One of the best protective coatings against severe wear or seizure of most metals is the normal surface oxide film. Mild wear removes surface oxide, which is replenished by atmospheric oxygen, either direct in dry sliding or dissolved in the lubricant in lubricated sliding. Where the surface contacts are severe enough to penetrate the oxide film into the unoxidized metal substrate, severe wear and seizure readily occur. Any failure to renew the surface oxide will increase the risk of severe damage.

Anti-wear additives are used to reduce the rate of wear when mild wear is taking place. The rubbing conditions will also be mild, and re-oxidation will normally be adequate to ensure the long-term protection of the surfaces. Anti-wear additives are therefore relatively stable chemicals which will adsorb or chemisorb on the oxidized metal surfaces. The films so formed protect the bearing surfaces at low loads and sliding speeds, at the same time reducing friction because their shear stress is lower than that of the metal or its oxide.

Typical anti-wear additives include the friction-reducing additives mentioned previously but not the friction modifiers. They also include slightly more polar and more reactive chemicals, typical examples being long-chain fatty acids such as stearic acid and palmitic acid, and their esters, such as ethyl stearate.

All the above substances are limited in their resistance to shear and their resistance to increased temperatures. A more powerful class of anti-wear additives has therefore been developed which consists almost entirely of phosphorus compounds. These include some of the best-known of all lubricant additives, such as TCP (tricresyl phosphate), TXP (trixylyl phosphate), ZDDP (zinc dialkyl dithiophosphate), and dilauryl phosphate.

If the sliding conditions are too severe, mild anti-wear additives will be inadequate to prevent removal of the oxide film, and consequent severe wear or seizure. More highly-reactive EP additives are then needed, and a number of the most common EP additives are listed in Table 3. It will be noticed that the table includes some compounds such as TCP and ZDDP which were previously mentioned as anti-wear additives. These compounds are particularly useful because they can act as anti-wear additives under mild conditions and as EP additives if the conditions become more severe.

The compounds in Table 3 are listed roughly in order of increasing EP activity, from the relatively mild phosphorus compounds to the active chlorine compounds. To some extent their activity is related to chemical reactivity and therefore to chemical instability. Because they are

Table 3 Some successful extreme pressure additives

Dilauryl phosphate
Didodecyl phosphite
Tricresyl phosphate (TCP)
Zinc dialkyldithiophosphates (Z/DDP)
Phosphosulphurized fatty oils
Zinc dialkyl dithiocarbamate
Mercaptobenzothiazole
Sulphurized fatty oils
Sulphurized terpenes
Sulphurized oleic acid
Alkyl and aryl polysulphides
Sulphurized sperm oil
Sulphurized mineral oil
Sulphur chloride treated fatty oils
Chlornaphtha xanthate
Cetyl chloride
Chlorinated paraffinic oils
Chlorinated paraffin wax sulphides
Chlorinated paraffin wax

designed to react with metal surfaces, especially at high temperatures, they all need some degree of thermal instability.

It follows that all the established EP additives have limited lives when exposed continuously to high temperatures, and this problem is discussed in more detail later. One of the most stable thermally is TCP, but even this has a very limited life at a lubricant bulk temperature of over 250°C, especially in contact with metals.

Similar thermal stability limits apply to most of the boundary lubricant additives in all four classes. Unsaturated molecules such as oleic acid and its derivatives, and the highly sulphurized materials, decompose at rather lower temperatures, while some of the simpler molecules such as lauryl alcohol, cetyl chloride and stearic acid will survive at 200°C only in the absence of oxygen.

Under oxidizing conditions the stability of the various phosphorus compounds is apparently not much reduced from their thermal stability limits, and ZDDP is a very effective anti-oxidant. Most of the other boundary additives have long-term temperature limits between 150°C and 230°C.

Because of concern about environmental pollution, pressure is growing for the elimination of phosphorus from lubricants. This pressure seems hard to justify, since the quantities used in lubricants are trivial compared with the amount of phosphorus used in agriculture or in the natural environment. Development of alternative classes of additive is likely to prove very difficult and costly.

6.3 ADSORPTION

Solid surfaces in contact with a liquid or gas will be covered with an adsorbed film of molecules from their environment. The adsorbed molecules lower the free energy of the surface, and the greater the reduction in free energy, the greater will be the stability of the adsorbed film. A typical adsorbed film, lauryl alcohol, is shown diagrammatically in Fig. 17.

It follows that some substances will be more strongly adsorbed than others, and on metal surfaces polar molecules will generally be more strongly adsorbed than non-polar molecules. In a mineral oil lubricant the molecules of the mineral oil itself are largely non-polar, while many of the additives used will be more or less polar. The additives will therefore

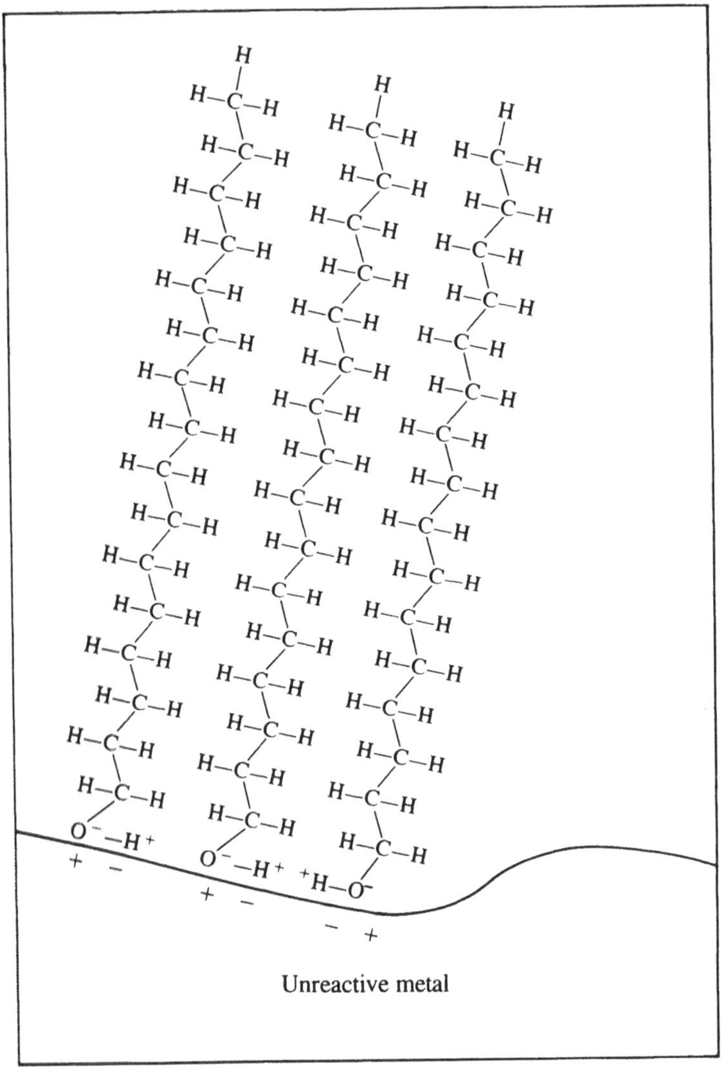

Fig. 17 Adsorption of lauryl alcohol

preferentially adsorb onto metal surfaces, and there may be competition for adsorption between different additives.

The adsorption process is reversible, and under equilibrium conditions there will be a balance between the rate of adsorption and the rate of desorption of a substance. The rate of adsorption is proportional to the concentration of the particular adsorbate in contact with the surface of the adsorbent solid. Similar balances will be set up for all the different molecular species present, so that in a complex mixture of additives the adsorbed film will include several of them. It is therefore important in formulating lubricants to ensure that additives which are intended to protect surfaces can compete for adsorption sites. The adsorbed film may be one molecule or more than one molecule thick.

Adsorption is important in boundary lubrication because friction-reducing additives, friction modifiers, anti-wear additives and even the milder EP additives are believed to act by adsorption. Effective friction-reducing and anti-wear additives usually have a slightly polar group which adsorbs onto the surface, and a long hydrocarbon chain which orients itself perpendicular to the surface to form a thick protective film. In addition adsorption may assist the action of other EP additives by concentrating them on a bearing surface and thus increasing their availability when required.

Adsorption is temperature-dependent and pressure-dependent. The amount of a substance adsorbed on a surface decreases as the temperature rises. It has been suggested that the effectiveness of adsorbed boundary additives ceases when the temperature reaches a specific desorption temperature. In some cases a marked increase in friction has been shown to occur when the temperature reaches the normal melting point of the adsorbate.

Where mixtures of additives are present in solution in a lubricant, any mutual solubility effects of additives or base fluid will depress individual melting points, so that there is no obvious reason for any correlation of desorption and melting point. If the correlation of desorption with melting point was generally true, most mild anti-wear and anti-friction additives would be rather ineffective, since most of them are liquid below 100°C. In practice most of them seem to be effective to temperatures of 120–150°C.

The important point is that the adsorption of all substances becomes less effective as the temperature rises.

Since adsorption is a universal phenomenon, it follows that all additives present in a lubricant will compete for adsorption sites on the bearing surfaces. The proportion of each additive adsorbed will depend on the polarity, the concentration, the critical desorption temperature, and, presumably, the melting point. Thus, even those anti-wear and EP additives whose effectiveness is considered to be due to chemisorption or chemical reaction are likely to be adsorbed at low temperatures.

The forces present between a surface and an adsorbed film are the weaker secondary electrical attractions known as van der Waals forces. Adsorbed films, therefore, have relatively poor resistance to mechanical shear, as well as to frictional heating.

6.4 CHEMISORPTION

Adsorption is a purely physical phenomenon, in which no chemical bonding, that is, exchange or sharing of electrons, takes place. Conversely, if any exchange or sharing of orbital electrons takes place between a surface and an adsorbed molecule, this represents the formation of a chemical bond, and the process is known as chemisorption.

The formation of a chemical bond represents a much stronger link than the Van der Waals forces present in adsorption, and chemisorbed films are much more resistant to mechanical shear, as well as to heat.

Carboxylic acids are typical of compounds which chemisorb onto surfaces, and Fig. 18 shows, diagrammatically, the chemisorption of stearic acid onto a steel surface. The combination of one of the iron atoms in the steel surface with the remainder of a stearic acid molecule after the loss of one hydrogen atom has created an iron salt of stearic acid, iron stearate.

Salts such as iron stearate are sometimes called soaps, although strictly speaking soaps are only the alkali or alkalyne-earth metal salts (e.g., sodium, lithium, calcium, barium). They do not have a single clear melting point, but go through a softening stage as the temperature rises. There is some evidence that a rise in friction occurs with chemisorbed films, as the temperature rises, at a temperature which coincides with a softening temperature of the relevant salt.

It can be seen from Fig. 18 that chemisorbed stearic acid has a structure similar to the adsorbed lauryl alcohol in Fig. 17, having a reacted group at the surface and a long carbon chain oriented perpendicular to the surface to form a thick film.

However, many of the organic phosphates, phosphites, and thiophosphates such as TCP and ZDDP are also believed to chemisorb onto bearing surfaces. Some of these compounds are effective anti-wear additives, even when they have no long carbon chains. The presence of a long perpendicular molecule is, therefore, not essential for surface protection provided that the chemisorbed film has a lower shear stress than the bearing surface itself and is strong enough to provide adequate resistance to shear.

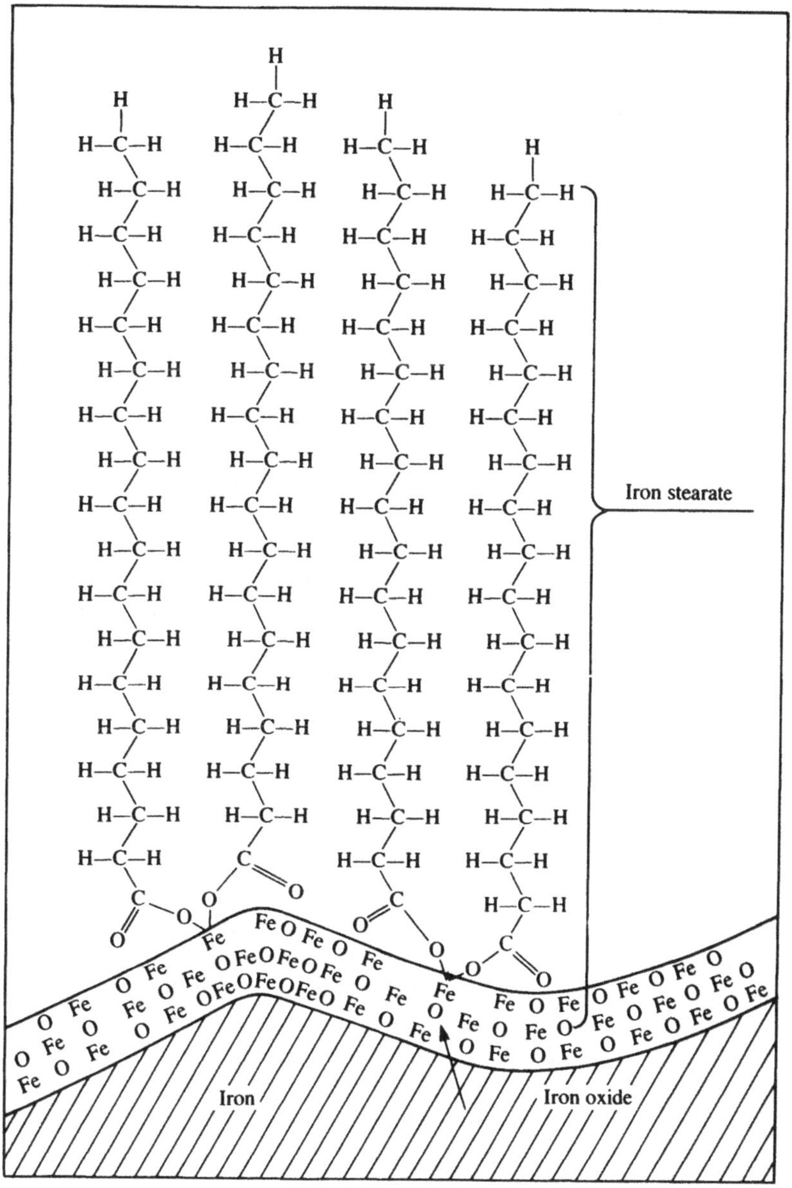

Fig. 18 The mechanism of chemisorption

Although some chemisorbed films show increases in friction as the temperature rises, the ultimate temperature limit for most chemisorbed substances is likely to be the temperature at which disruption of the molecule begins.

For carboxylic acid soaps this temperature is probably in the region of 200°C. For phosphate esters it is likely to be only a little higher. Thus, although chemisorbed substances have better mechanical and thermal resistance than adsorbed films, their suitability for high temperatures is still quite limited.

6.5 CHEMICAL REACTION

The benefit of adsorbed and chemisorbed additives is limited to friction control and to reducing mild wear, where the protective oxide film on a bearing surface is not seriously disrupted. If the oxide film is removed from the surface and leaves bare metal, compounds such as hydrocarbons, alcohols, and fatty acids have been shown to decompose in contact with the surface to give small fragments, mainly gaseous, which have no value for surface protection.

The value of oxygen itself as a boundary lubricant has already been mentioned. When sliding speed and load are high enough to cause removal of the oxide film from a surface, immediate re-oxidation will re-establish protection and prevent severe adhesion, galling, and scuffing or scoring.

The difficulty is that oxygen is only present at a lubricated bearing surface as gas dissolved in the lubricant. Its availability is limited by its absolute solubility and its rate of solution in the lubricant, as well as by the rate at which it can diffuse through the lubricant. The availability of oxygen can be increased by operating a system with an atmosphere of pure oxygen or by dissolving oxidants in the lubricant. This will improve the surface protection, but is counter-productive, because the rate of oxidation of the lubricant itself is also increased.

A very much better solution is to use sulphur instead of oxygen. Sulphur is much more soluble in most lubricants, and being a solid at normal temperatures, it does not evaporate from the lubricant surface. It is also much less reactive than oxygen, and does not cause degradation of lubricant base oils or most additives.

When severe sliding causes removal of the oxide film from a metal surface, dissolved sulphur will rapidly react with the freshly-exposed metal to form a metal sulphide coating on the surface. The surface sulphide provides a physical barrier against the sliding counterface, has lower shear stress than the metal substrate, and has little tendency to adhere to the counterface. Sulphur can also be provided in the form of sulphur compounds, several of which are listed in Table 3.

Phosphorus and chlorine are also effective in forming protective coatings of metal phosphide or chloride, respectively, on exposed metal surfaces. Neither phosphorus nor chlorine can be provided in uncombined form because of their high toxicity and other hazards. They are always used in the form of compounds, several of which are listed in Table 3. As a general rule phosphorus compounds are less reactive than sulphur and the sulphur compounds, and chlorine compounds are the most reactive of the three groups.

All these EP additives are designed to be most effective when incontact with metal surfaces which have lost their surface oxide due to severe sliding against a counterface. The highly active bare metal surface causes complete disruption of the additive molecule and liberates sulphur, phosphorus, or chlorine either as a free element or combined in a small molecule or radical, and this fragment reacts rapidly with the bare metal to form a coating of metal phosphide, phosphate, sulphide, sulphate, or chloride.

The reaction of any of these substances with the metal is, in fact, a form of corrosion, and the combined effect of reaction and wear represents corrosive wear. The selection of an EP additive is a compromise between desirable EP protection and undesirable corrosion. The performance of the additives is, in fact, helped by the temperatures present.

An asperity collision on the sliding surfaces which is strong enough to remove the surface oxide will also cause a high flash temperature. If we use the guideline mentioned earlier, that the rate of a reaction doubles for a 10°C rise in temperature, then a flash temperature of 200°C would increase the reaction rate between additive and surface by a factor of 10^{20}, or over a million. This relationship will generally be assisted by the fact that attack by the additive on the normal oxidized metal surface will be much less than its attack on the local bare spots.

An EP additive can therefore be selected to give an acceptably low rate of attack on the oxidized metal surface at bulk temperatures, but an effectively high rate of attack on a bare metal surface at flash temperatures.

Surface coatings of metal phosphide or sulphide can be oxidized under suitable conditions of high temperature and oxygen availability. The products will be phosphate or sulphate. However, if sufficient oxygen is present for these reactions, there will also be enough to re-oxidize the bearing surfaces, and the need for EP additives will be less. In any case, the resulting metal phosphate or sulphate will also protect the sliding surfaces.

In the absence of significant amounts of oxygen, a coating of metal phosphide, phosphate, sulphide, sulphate, or chloride has great thermal stability, and will provide protection to temperatures between 600 and 1200°C for different metals. Different coating compounds reach their melting points at temperatures within this range, but it is believed that this improves their beneficial effect on lubrication.

The problem with existing EP additives lies in their instability and reactivity as the temperature of the bulk of the lubricant rises. At some temperature between 150 and 250°C most current EP additives will begin to decompose rapidly. Their decomposition products, and the few additives such as sulphurized mineral oils and cetyl chloride which remain relatively stable, all become increasingly aggressive to the metal surfaces.

The result is that all present EP additives become unusable as bulk temperatures approach 260°C. In this context bulk temperature means all temperatures other than flash temperatures, and includes the temperature which the bulk of the oil reaches as it passes through a bearing, gear, or other component.

The search for boundary additives for high-temperature use has so far not been very successful. From time to time success has been reported for various substances, such as triphenyl antimony sulphide, and a chlorinated tri-(phenoxyphenyl) phosphate for about 280°C, but none is known to have gone into commercial use.

Reactive EP additives are often used in conjunction with anti-wear or mild EP additives of the adsorbing type, and the combination is often found to give a marked improvement in lubrication. The improvement is believed to be due to very efficient adsorption of additives such as the long-chain alcohols, acids, and esters on a surface coating of reacted metal phosphide, sulphide, or chloride. This effect is likely to be particularly useful where the occurrence of oxide removal is only occasional or takes place only in a particular phase of operation, such as during running-in.

The combined effect of adsorbed and reacted films will not assist in higher-temperature operations because the adsorbed film will generally cease to be effective before the temperature limit of the reactive EP additive is reached.

6.6 POLYMERIZATION

Under certain circumstances a coating of polymeric material, called friction polymer, can be formed on a bearing surface. The mechanism by which the polymerization takes place seems fairly clear and consistent. The exposure of fresh metal, denuded of its surface oxide by wear, has been stated earlier to lead to high surface energy and catalytic effect. These effects can cause disruption of organic molecules in contact with the surface, and one result of molecular disruption is likely to be the creation of free radicals.

Under conditions which have not been clearly established, the reactive free radicals lead to addition reactions with other molecules, and this type of addition can propagate polymerization.

The conditions which encourage the formation of friction polymer seem to include the maintenance of a generally high surface temperature with sufficiently severe sliding to cause continuing wear, but where the severity of the situation is insufficient to cause scuffing. It is possible that the presence of wear debris is essential to provide an additional surface area of active metal to catalyse the polymerization.

Analysis of friction polymer formed in a mineral oil containing EP additives shows the presence of a high proportion of carbon, the active elements of the EP additives, and metallic elements representative of the bearing surfaces.

In the particular case of certain phosphorus-containing additives in mineral oil, a thick film of polymerized material has been shown to be formed in pure rolling contact at temperatures between 100 and 170°C. The product can have a thickness anywhere in the range 0.005–0.5 μm. The thicker films tend to be soft and easily removed in sliding, but it has been postulated that their structure helps to provide a viscosity increase which is helpful under EHD conditions. The structure of the polymer consists of a three-dimensional phosphorus-oxygen framework with attached alkyl groups.

Silicones also produce a three-dimensional polymeric structure in which the framework consists of cross-linked silicon-oxygen chains.

In all cases the polymers provide some surface protection and are more temperature stable than the original substances present, but no details of their thermal and mechanical stability have yet been published.

The formation of friction polymers appears to offer one of the few routes to achieving improved high-temperature lubrication, but so far there has been no evidence of any major effort to develop this possibility.

6.7 MICRO-RHEODYNAMIC LUBRICATION

When two asperities on sliding surfaces come into contact in the presence of a liquid lubricant, the situation is analogous to the lubricated contact between components in rolling bearings or gears. For this reason it is considered probable that micro-elastohydrodynamic lubrication (Micro-EHD) can take place between the interacting asperities in the same way as EHD lubrication between non-conformal contacts.

The situation on the micro-scale is, however, relatively much more complex than on the macro-scale. Because of the short time-scale of asperity interactions the transient thermal gradients are likely to be high, with corresponding effects on the entry zone viscosity. There may also be local thickening effects due to adsorbed or polymeric films. The shear rate in the lubricant film will often be much greater than in rolling bearings or most gears. According to the Rehbinder theory, the yield stress will be reduced by the presence of adsorbed films, and there will often be plastic flow of the asperity material, as well as elastic deformation. In addition, the oxide or EP films on a metal surface will have a proportionately greater effect in creating non-uniform material properties in the solid surface.

Because of the occurrence of shear and flow, not only in the liquid lubricant but also in the adsorbed, reacted, or polymeric films and in the solid material of the asperity, the phenomenon is called micro-rheodynamic lubrication.

There is little or no firm evidence for the contribution of micro-rheodynamic lubrication to the overall behaviour in boundary lubrication, but there is now a general acceptance that it probably makes a significant contribution, at least in and near the mixed lubrication regime.

Almost every factor involved in micro-rheodynamic lubrication would be affected by increased temperature. The effects would include the following.

(1) Decrease in bulk lubricant viscosity (detrimental).
(2) Decrease in pressure coefficient of viscosity (detrimental).
(3) Reduction of adsorbed films (detrimental).
(4) Possible increase in polymeric films (advantageous).
(5) Reduction in elastic modulus (probably advantageous).
(6) Reduction in yield stress (probably advantageous).

CHAPTER SEVEN

Liquid Lubricants

7.1 GENERAL

Lubrication by liquids has several important advantages over grease or solid lubrication. Lubricating oils can be used over a far wider range of speeds than greases or solid lubricants, and for the same load-bearing performance as greases. The fluidity of liquids enables them to provide much more effective cooling than greases or solid lubricants, either by convection or by forced circulation. The same fluidity makes it much easier to re-supply oils to bearings, and these last two factors are particularly important in high-temperature applications.

It follows that liquid lubricants can conveniently be considered as the first choice of lubricant, with which alternative types can be compared, and the temperature limits of liquid lubricants will therefore be described first. The term 'liquid lubricants' is more general than 'oils' because some of the liquids available for high-temperature applications are very different from the usual image of an oil, but 'lubricating oils' will be used for convenience.

The main factors which determine the maximum temperature for the use of a particular lubricant have already been described. These are thermal degradation, oxidation, reduced viscosity, and volatility, as well as the loss of effectiveness of anti-wear, EP, and other additives. In this chapter these limitations will be considered in more detail with respect to specific types of lubricating oil.

Thermal degradation and oxidation are time-dependent, so that for those factors the limiting temperature depends on the required life. Any quoted temperature limit must, therefore, be very arbitrary unless it is related to the required life and the degree of deterioration which is considered acceptable. For this reason the temperature limits given for individual lubricant types will be presented in the form of graphs of life against temperature.

It is instructive to consider the proportion of the total operating life which an oil spends at different temperatures. As an example, the case will be considered of an oil subjected to the four heating regimes

Table 4 Times of exposure of oil to different temperature regimes

	Recirculating systems	*Replenishment systems*	*Non-replenished systems*
Ambient temperature	10^5–10^8 s		
System temperature	10^4–10^7 s	10^3–10^4 s	
Component temperature	10^3–10^4 s	10^4–10^6 s	10^5–10^7 s
Flash temperature	10^{-4}–10^{-1} s	10^{-3}–1 s	10^{-2}–10 s

discussed in Chapter Two, namely ambient temperatures, system inputs, component inputs, and flash temperatures.

It is possible to define broadly the proportions of total time which the oil in a system spends at each of the four resulting temperatures. Table 4 gives some estimates of those times for three types of system, namely recirculating systems, replenished systems without recirculation, and non-replenished (i.e., self-contained) systems. In each case the times relate to one year of continuous operation.

The times in Table 4 are, of course, only rough estimates, but if they are correct they indicate that the oil in self-contained systems is exposed to component and flash temperatures ten times as long as in replenished systems and one hundred times as long as in recirculating systems.

It was shown previously that the rate of oxidation is relatively slow and uniform until the anti-oxidant is fully depleted, and then becomes rapid. In the case of thermal degradation, the rate is fairly steady at a uniform temperature. On the basis of the guideline used previously, that the reaction rate doubles for a 10°C rise in temperature, then from the above figures it follows that for a particular level of deterioration recirculating systems can be used for temperatures 30–40°C higher than replenishment systems and 60–70°C higher than self-contained systems.

If we consider a hypothetical recirculating system in which the reservoir temperature is 30°C, system temperature 50°C, component temperature 80°C, and flash temperature 280°C, then we can calculate the proportion of the total deterioration (e.g., thermal degradation or anti-oxidant depletion) which occurs in each of the four regions.

These figures are shown in Table 5. They are, of course, entirely hypothetical and over-simplified, but the results are probably not unrealistic. They show that, even with a fairly high flash temperature of 200°C above the bulk temperature, the contribution of the flash temperature to the overall oil degradation is very small. This will not be equally

Table 5 Oil decomposition in hypothetical system

	Temperature (°C)	*Time* (s)	*Relative reaction rate*	*Total reaction* (%)
Reservoir temperature	30	10^6	R	57
System temperature	50	10^5	4R	23
Component temperature	80	10^4	32R	18
Flash temperature	280	10^{-3}	3.4×10^7R	2

true in a self-contained system, but even there the deterioration due to flash temperatures would not be more than 10 per cent of the total. (This conclusion will not be true for grease or solid lubrication, and this will be discussed again in Chapters Eight and Nine.)

Table 5 also shows that, although the assumed reservoir temperature is very low, because of the long residence time in the reservoir its contribution to the overall deterioration of the lubricant is very considerable. It is therefore particularly important in a recirculating system to ensure that the reservoir temperature is kept as low as possible, quite apart from its influence in reducing the feed temperature to all the other components.

The life of an oil at a particular temperature also depends on the amount of degradation which is acceptable, and this, in turn, depends on how much the performance can be allowed to deteriorate. This will vary with the specific oil and with the application. Any indication of a maximum temperature and life will again be quite arbitrary, and the eventual decision about both must be made by the user. Information will therefore be given on the nature of the degradation and its effects on performance, so that the user can relate them to his own equipment.

The eventual failure of a machine due to lubricant deterioration will often be indicated by an exponential rate of change in one or several criteria, such as temperature, vibration, corrosion, loss of precision, or loss of power. Even then the decision will often be subjective as to when an actual state of failure has occurred or the machine should be stopped.

For example, the degradation of a mineral oil due to oxidation causes acidity, increase in viscosity, and sludging. These will, respectively, cause corrosion, reduction in oil flow, bearing wear, and overheating, and eventually oil blockage. It would normally be expected that the machine would fail by seizure, especially in the case of an internal combustion engine.

However, in one series of experiments, the decision was made to allow a diesel engine to operate continuously until it failed. The oil fill was kept topped up, and fuel was supplied, but no other maintenance was carried out. The decision was finally made to stop the test when the vibration became so bad that there was a real risk that the engine would break loose from the test bed. At that time there was some overheating, but very little reduction in power. After shutdown there was found to be considerable wear and some bearing corrosion, and on cooling the oil solidified due to oxidation and sooty carbon.

There are two relevant points about this experience. The first is that the oil was so badly degraded that if the engine had been in intermittent operation it would much earlier have reached a point where once stopped it could not have been started again. It might in fact have been described as having seized. The second point is that the nature of the eventual 'failure' was not what would have been predicted from the nature of the oil degradation.

This case shows clearly the difficulty of defining the stage at which the condition of an oil becomes unacceptable, and thus of predicting the usable life of an oil.

Some authorities have defined oil life as the time taken for 10 per cent deterioration, but this is far too simplistic. A 10 per cent change in viscosity can be unimportant or even trivial, whereas a 10 per cent content of oxidation products is enormous. In the case of polyglycols, and some other synthetics, the degradation products are volatile and are lost from the system, so that deterioration means a loss of volume, but only a very, very slow loss of quality, and if relatively unimportant.

In many circumstances the life of an oil may be determined by depletion of additives, especially anti-oxidants, dispersants, or overbased calcium additives, leading to excessive oxidation, sludging, or acidity, respectively. The result is therefore not directly a function of the base oil at all.

It was said earlier that one survey had shown that an oil viscosity of 1cSt is the minimum for any satisfactory lubrication, and this will be used as a guideline in this chapter. However, there will be some bearings and even gears that will operate satisfactorily with lower viscosities, especially at low loads, and there will be others that require much higher viscosities.

This minimum oil viscosity is defined on the basis of the viscosity required for satisfactory hydrodynamic or elastohydrodynamic lubrication. With externally-pressurized ('hydrostatic') lubrication, the fluid viscosity is much less important, and even gases can be used to lubricate quite heavily-loaded bearings.

The temperature limit graphs in the remainder of this chapter must therefore be considered only as guidelines, which in some cases may be in error by 50°C or more. They are subject to all the uncertainties which have been mentioned previously, namely the effects of different anti-oxidants, the effect of the availability of oxygen, the catalytic effect of metal, the nature of the deterioration, and the acceptable extent of deterioration.

The figures also suffer from one further disadvantage. They are all based on published data. In all cases the data come from a variety of different publications, using different experimental techniques and different criteria of deterioration. Even where the methods and criteria were the same, the reproducibility of the results was often poor.

It follows that a great deal of subjective judgment had to be used in collating the data and preparing the graphs, and they must be recognized as very imprecise. They are intended to be slightly conservative, but the performance of an oil is dominantly influenced by the conditions in a particular machine. Any user with a high temperature application should carry out his own tests to ensure the suitability of an oil for his particular application.

One final caution is to beware of the word 'synthetic'. Most of the oils described in this chapter are synthetic, but they vary enormously in their behaviour. Synthetic hydrocarbons resemble mineral oils rather than other synthetics, and are so different from the synthetic perfluoropolyalkyl ethers and silicones that they should not be spoken of in the same breath. If people mention a synthetic oil, always find out what type of synthetic oil they are talking about.

7.2 MINERAL OILS

There is a wide variety of lubricating oils which consist basically of hydrocarbons, that is chemical compounds containing only carbon and hydrogen. They range from the lowest quality naphthenic oils to pure synthetic poly-olefines, and all of them may sometimes be referred to as mineral oils.

To be completely pedantic, minerals are materials obtained by mining, so that mineral oils would properly refer only to oils obtained from natural petroleum deposits, with a greater or lesser degree of subsequent refining. On this basis hydrocarbon oils can be conveniently divided into mineral oils and synthetic hydrocarbons, and this classification will be used in this chapter.

However, by extremely severe refining processes, oils from petroleum can be produced whose properties are much closer to those of synthetic hydrocarbons than they are to those of ordinary mineral oils. These include the Deep Dewaxed Super-Refined mineral oils of the 1960s, and the Very High Viscosity-Index (VHVI) and Ultra-High Viscosity Index (UHVI) products of the 1990s.

For convenience the hydrocarbon oils will therefore be divided into two sections. In this section the title 'Mineral Oils' is used to describe the oils obtained by conventional refining processes (by the standards of 1993). In the next section the title 'Synthetic Hydrocarbons' will be used to include the true synthetic hydrocarbons, Deep Dewaxed Super-Refined mineral oils, and VHVI and UHVI products.

It is not proposed to describe the chemical nature of mineral oils in much detail, but some knowledge of the chemical types present is necessary for an understanding of their behaviour at high temperature.

Mineral oils consist mainly of hundreds of different hydrocarbons, together with small quantities of compounds which contain oxygen, nitrogen, or sulphur, and sometimes other elements, in their molecules. The hydrocarbons themselves are classified as three distinct types, paraffins, naphthenes, and aromatics, whose structures are shown in Fig. 19. Each carbon atom has four bonds, and each hydrogen one bond. The paraffins contain only single bonds, either C—C or C—H. The carbon atoms are joined together in chains, which are either completely straight in normal paraffins (n-paraffins) or branched (in iso-paraffins). The naphthenes also contain only single bonds, but some of the carbons are joined together in rings which consist almost invariably of five or six carbons. The aromatics also contain rings, but those rings include double bonds.

The situation is further complicated by the fact that a naphthenic or aromatic ring may be attached to a long paraffinic chain. In general the character of the oil is determined by the proportion of the total carbon present which is in the form of paraffinic chains, asphaltic rings, or aromatic rings. An oil would be described as naphthenic if more than about 40 per cent of the carbon is in naphthenic rings; otherwise it would be called paraffinic.

The compounds containing oxygen, sulphur, nitrogen, or other ele-

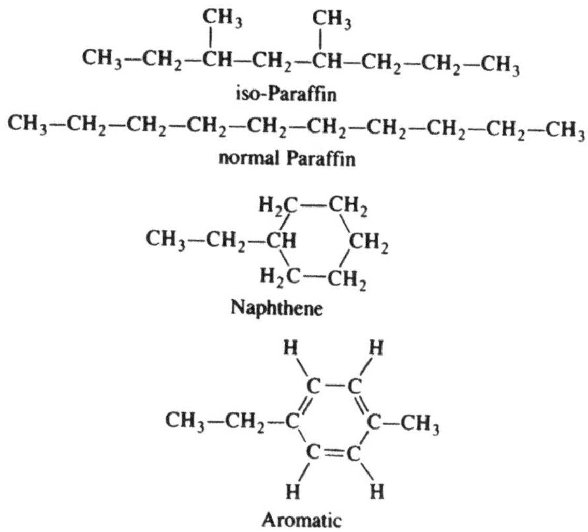

Fig. 19 Basic hydrocarbon types

ments other than carbon and hydrogen are sometimes called asphaltenes when present in crude oil or the heavier lubricating oils, although technically asphaltenes are the heavier, less soluble molecules. These compounds and the aromatics generally lead to chemical instability, and most of them are removed by the standard refining processes. Paraffinic oils also contain wax, which is a mixture of high molecular weight paraffinic hydrocarbons. Wax is undesirable for low-temperature performance and is also a valuable by-product, so paraffinic oils are normally dewaxed.

Naphthenic oils are useful for some purposes, but they have low viscosity indices and their thermal stability is inferior to that of paraffinic oils. The best mineral oils for high-temperature applications are paraffinic oils which have been highly refined and dewaxed. It is now normal practice to refine the base oils severely to remove almost all traces of unstable molecules and sulphur. This leaves a product which has high viscosity index and good thermal stability, but poor oxidative stability and poor natural anti-wear properties. Fairly high concentrations of additives are therefore used to ensure satisfactory performance in every necessary respect.

The remainder of the information in this section will relate to the performance of mineral oils which are suitable for high-temperature use and have appropriate additive packages.

The thermal decomposition temperature for paraffinic hydrocarbons in accordance with the ASTM D2879 test (T_D) has been found to be about 350°C, which is comparable with that of the best polyol esters used in jet engines. The thermal decomposition temperature cannot be improved by additives, but can be reduced by the presence of less stable compounds, including certain additives. A more realistic range for good quality mineral oils would probably be about 280–320°C.

Significant oxidation takes place at much lower temperatures, and is strongly influenced by the presence of metals which can catalyse oxidation. It was explained in Chapter Five that oxidation takes place relatively slowly while anti-oxidants are present, but at much higher rates once the anti-oxidant is depleted.

Both thermal degradation and oxidation are time-dependent, and Fig. 20 gives approximate indications of the life of good quality mineral oils as determined by thermal stability in the absence of oxidation, and by oxidation with and without anti-oxidants. The limits have been drawn as broad bands instead of lines because of the uncertainties mentioned previously.

The nature of the degradation products has been described previously, but can be summed up as acidity, increase in viscosity, and, eventually, sludging.

The viscosity index of conventional highly-refined paraffinic oils lies generally in the range 80–110. The actual viscosity depends on the average molecular weight of the various hydrocarbons present and on the range of molecular weights. The variety of mineral oils available is so great that it should be possible to obtain an oil with adequate viscosity at any temperature up to the maximum practical limit.

Figure 21 shows the variation of viscosity with temperature for a series of mineral oils having viscosity index of 100. It can be seen that the most viscous oils have useful viscosities at very high temperatures, but that their viscosities at low temperatures are so high that they are likely to cause handling difficulties. They will also cause high viscous friction at low temperatures, as shown in Fig. 14.

The volatility of a mineral oil depends on the molecular weight of the smallest molecules present. During the manufacture of mineral oils the crude oil is distilled under vacuum. The more viscous lubricating oils are manufactured from the residue remaining after two distillation processes,

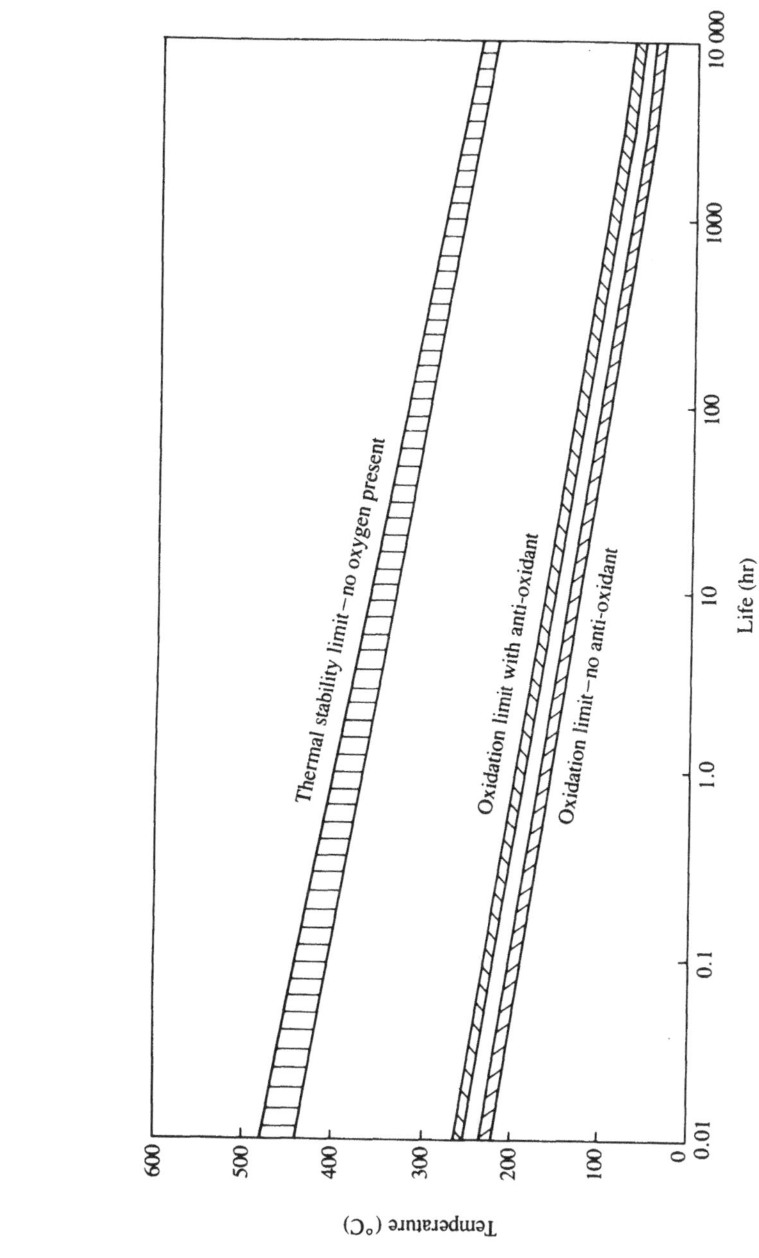

Fig. 20 Temperature limits for mineral oils

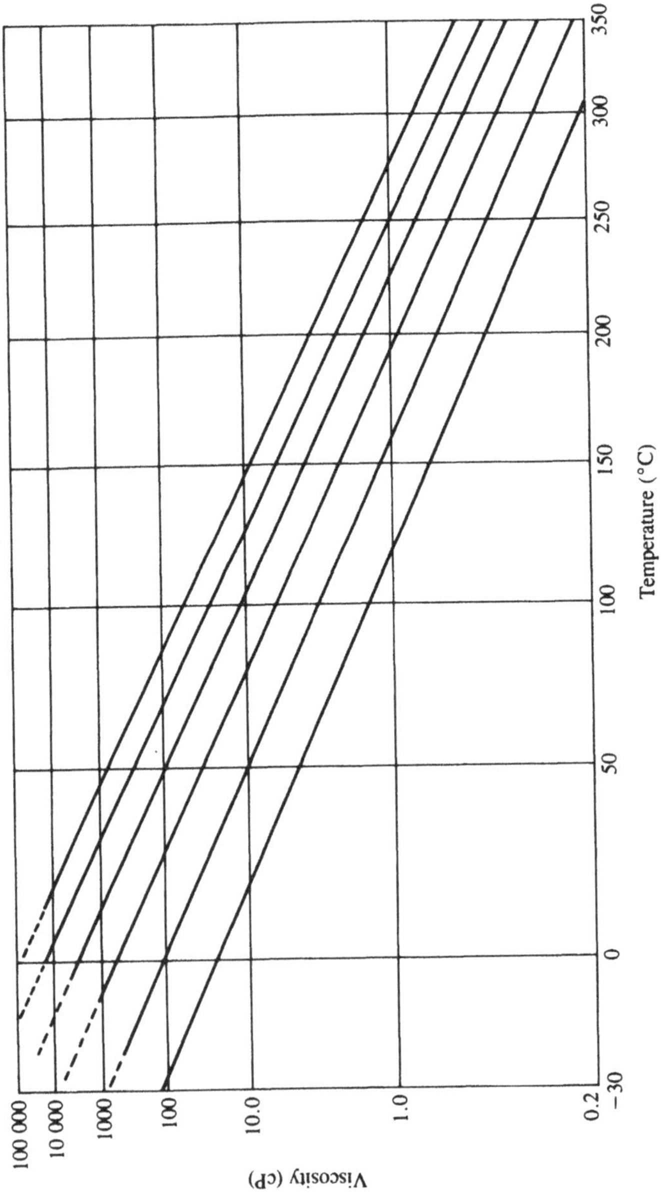

Fig. 21 Variation of viscosity with temperature for mineral oils

first at atmospheric pressure and second under vacuum. It follows that the volatility of any remaining hydrocarbons is very low.

In any high-temperature application there will be some loss of oil by evaporation, such as in the cylinders of internal combustion engines. The actual proportion of the oil lost by evaporation will depend on the temperature, pressure, and degree of ventilation. Where oxygen is present it can usually be assumed that at any temperature the loss by evaporation will be less than the loss by oxidation. Where oxygen is not present, the features which keep the system inert, such as sealing or inert gas pressurization, will also tend to reduce evaporation. It is only in vacuum at temperatures above 250°C that evaporative loss is likely to be high enough to cause lubrication problems.

One problem which needs to be considered when any liquid lubricant is used at high temperature is that of ignition or flammability. The tendency of a liquid to ignite depends on its volatility and molecular composition, but also on many extraneous factors, such as the amount of oxygen present, the degree of splashing or spraying of the liquid, the pressure, and the presence of sparks or flames. As a result there are a variety of different tests simulating different situations, and it can be quite difficult to compare the flammability of different lubricants.

Mineral oils are generally flammable. The flash point, at which the vapour can be ignited by a small flame, tends to be about 200°C or higher. The autogenous or spontaneous ignition temperature, at which an oil ignites if injected into a hot vessel containing air, is about 350–380°C. However, ignition can be produced more easily if a fine spray of mineral oil impacts onto a hot surface at temperatures above about 250°C. Once a hot mineral oil is ignited, combustion is self-sustaining.

Mineral oils are good lubricants, but their anti-wear and load-carrying performance is limited unless improved by additives. Most additive technology has been developed for mineral oils, so the range of additives available is enormous. As a result the load-carrying properties can be improved to levels as high as for any type of lubricant.

7.3 SYNTHETIC HYDROCARBONS

The most important synthetic hydrocarbons for lubrication at present are almost completely paraffinic. They include polyalphaolefines (PAO) and polyisobutenes (PIB) which are both relatively low molecular weight

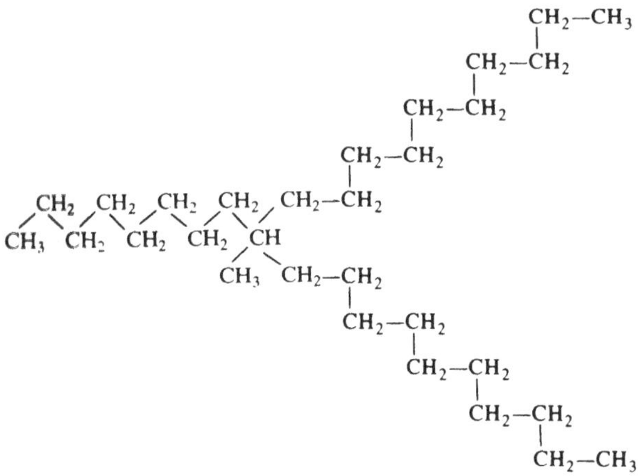

Fig. 22 Structure of a typical polyalphaolefine (α-decene trimer)

polymers and are truly synthetic. Their molecular weights lie in the range 250–500.

Very similar in properties are base oils produced from petroleum by a variety of processes which first remove all the unstable sulphur and nitrogen compounds. The next stages are called hydrocracking stages. They first hydrogenate the aromatic rings to naphthenic rings and then open up a high proportion of the naphthenic rings. The result is a mainly branched paraffinic structure of linked long straight chains with attached naphthenic rings. These structures are very similar, and in some cases almost identical, to that of the typical PAO shown in Fig. 22. An alternative process which produces similar base oil fractions is wax isomerization.

The base oils produced in this way are similar in behaviour to earlier oils known as Deep-Dewaxed Super-Refined (DDSR) mineral oils, but the newer hydrocracking processes are much less expensive.

Because of the similarity in the properties of these different oils it will be convenient to group them together as synthetic hydrocarbons.

The thermal degradation and oxidation mechanisms are similar to those of mineral oils, and were described in detail in Chapters Four and Five. They have better thermal stability and oxidation resistance, and this is reflected in the higher temperature limits shown in Fig. 23.

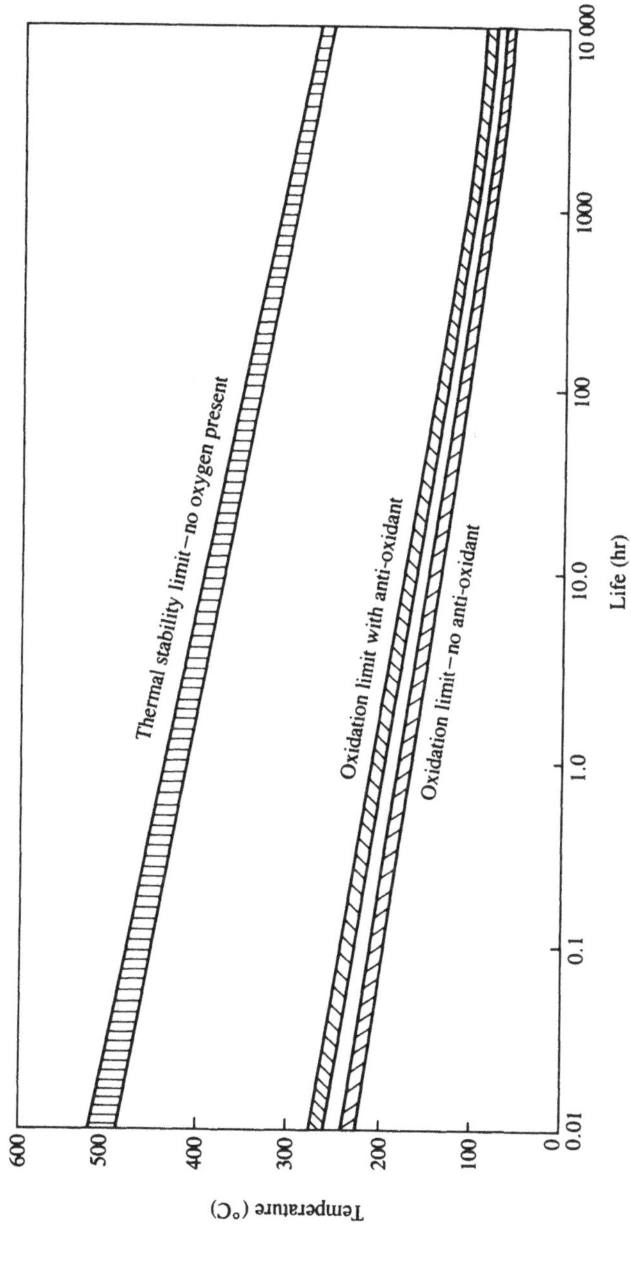

Fig. 23 Temperature limits for synthetic hydrocarbons

The limits at first glance are not much higher than those for the mineral oils, but for example the life of an inhibited synthetic hydrocarbon (i.e., one with an anti-oxidant) at 120°C would be roughly 500 hours, compared with 100 hours for a mineral oil. The advantage of the synthetic hydrocarbons would, in fact, be even greater because they have better response to anti-oxidants, giving longer life before anti-oxidant depletion.

The nature of the decomposition products is generally similar to that of the mineral oils, but there is an indication that the absence of aromatic compounds results in less tendency to form tarry deposits and coke.

The viscosity indices of the synthetic hydrocarbons are higher than those of mineral oils, being generally greater than 120. They are therefore known as Very High Viscosity Index (VHVI) oils, while those with even higher viscosity indices, greater than 140, are sometimes called Ultra-High Viscosity Index (UHVI) oils. Figure 24 shows the variation of viscosity with temperature for four typical synthetic hydrocarbons.

Synthetic hydrocarbons are now in widespread use for blending into base-stocks for vehicle engine oils, but the range of viscosity grades available is far more restricted than for mineral oils. Although it should theoretically be possible to produce synthetic hydrocarbons with adequate viscosity for use for short periods at temperatures up to 300°C, such oils may not be readily available at present.

The high viscosity indices available with these oils allow fully-formulated multi-grade oils to be produced with lower concentrations of viscosity-index improver. This results in a reduced risk of problems caused by temporary or permanent loss of viscosity due to shear.

In the United States a class of very stable hydrocarbons called Severely-Hydrogenated Oils is in commercial use for vehicle engine oil base-stocks. They are manufactured from a petroleum vacuum distillate by severe hydrogenation, but without significant cracking. The result is an oil which has thermal and oxidative stability similar to that of the synthetic hydrocarbons described above, but without the high viscosity indices.

Polyisobutenes (PIB) are synthetic hydrocarbons manufactured by polymerization of iso-butene. Their thermal and oxidative stability and viscosity index are not as high as those of the PAO and hydrocracked oils, but are better than those of conventional mineral oils. They are also better solvents for additives and have the major advantage of producing no deposits when they are thermally degraded or oxidized. They decompose by an 'unzipping' mechanism, giving low-molecular-weight products

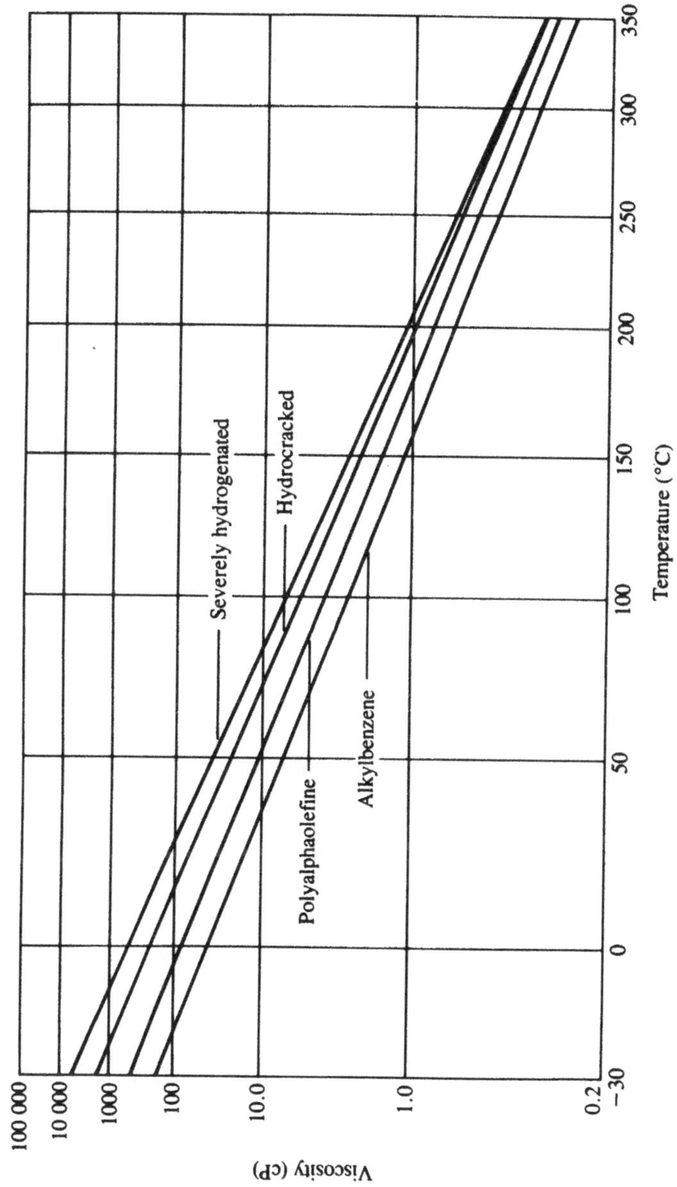

Fig. 24 Variation of viscosity with temperature for synthetic hydrocarbons

which volatilize, leaving a clean residual oil. Their major volume uses are in two-stroke engine oils and in certain automotive gear oils.

One final class of synthetic hydrocarbons is that of the alkylbenzenes. These are widely used commercially in the manufacture of detergents and over-based sulphonate engine oil additives. They have high viscosity indices between 100 and 130, and good thermal stability, but the commercial products have poor oxidation resistance. Special versions have been synthesized and evaluated for use in aircraft hydraulic systems, and these can be formulated to have satisfactory oxidation resistance, but their development is not yet very advanced. Alkylbenzenes have low viscosity at low temperatures, as shown in Fig. 24, and are used in refrigerator oils and Arctic engine oils.

The volatility of all these synthetic hydrocarbons is much lower in relation to their viscosity than that of the mineral oils. This is because they contain little or no volatile low-molecular-weight material.

It is probably for the same reason that they have better resistance to ignition than the mineral oils, although they must be considered generally as flammable substances.

In general the additive technology which has been developed for mineral oils is also suitable for synthetic hydrocarbons, and their lubricating performance is excellent. However, their solvent properties are not as good as those of mineral oils, so they may not be able to accept such high concentrations of additives. As a result, it is not yet clear whether they can reach the same level of EP properties as mineral oils.

7.4 DI-ESTERS

The group of synthetic lubricants commonly known as esters consists of a variety of alkyl esters of carboxylic acids. They fall conveniently into two classes, known as di-esters and 'hindered' or 'polyol' esters, and examples of both classes are shown in Fig. 25. The properties of the two classes differ enough for them to be described separately, and the di-esters will be covered here and the polyol esters in the next section.

The di-esters were the first class of synthetic oils to be used in large volume. The earliest aircraft jet engines were lubricated with mineral oils, but the temperatures were too high for satisfactory performance. Di-esters were found to be the best oils available for that application, and by about 1958 most aircraft gas turbines were lubricated with di-esters.

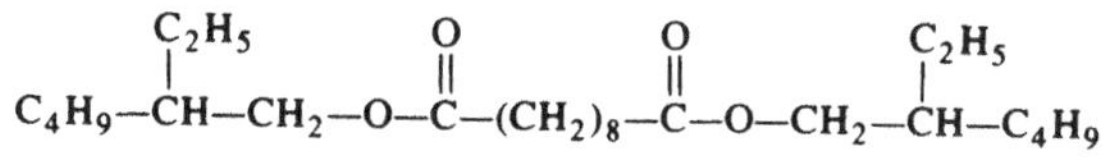

Typical di-ester (Di-2-ethylhexyl sebacate)

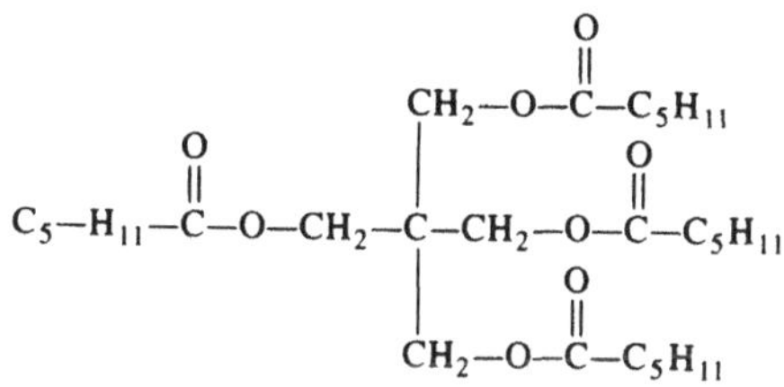

Pentaerythritol tetrahexanoate

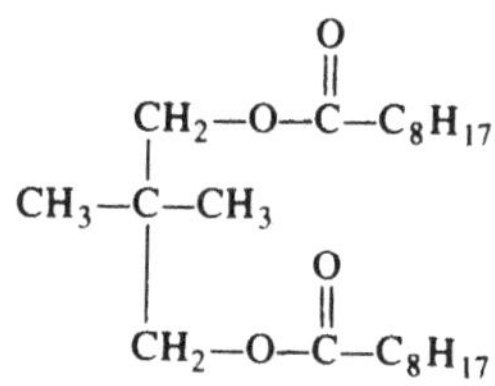

Neopentyl glycol dinonanoate

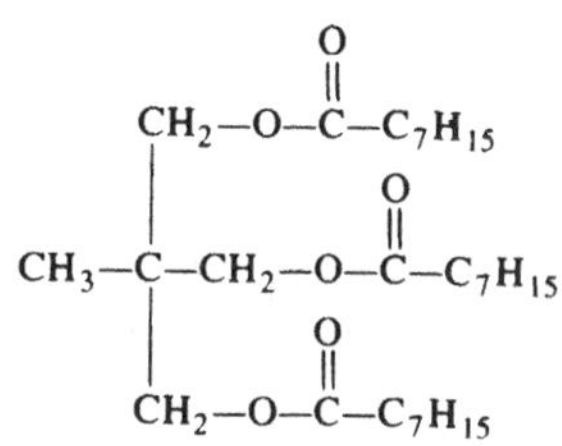

Trimethylolethane trioctanoate

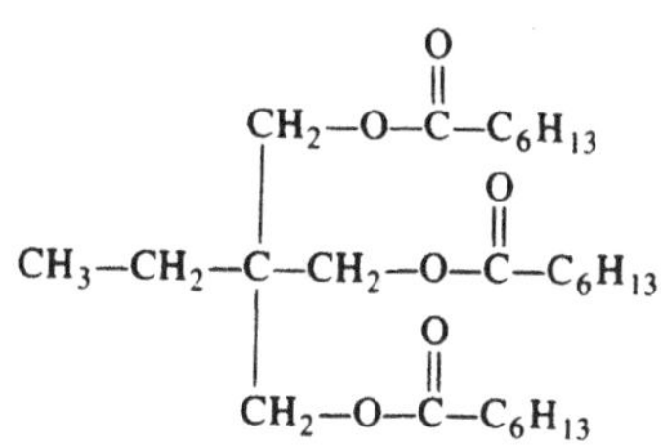

Trimethylolpropane triheptanoate

Fig. 25 Structure of typical di-ester and polyol esters

The thermal and oxidative degradation of the di-esters have been described in Chapters Four and Five. A typical initial thermal degradation temperature (T_D) is about 275°C, but in contact with some metals this is reduced to about 230°C. Approximate temperature limits for thermal and oxidative decomposition are shown in Fig. 26. The limits are drawn quite broadly because there is considerable variation between different esters. There is a compromise between the viscosity properties required and the thermal and oxidative stability which can be achieved. The decomposition products are generally similar to those of the synthetic hydrocarbons.

The most readily available grades of di-ester are those which are used for jet engine lubrication, although those may increasingly be blended with some polyol esters. The two standard grades of di-ester are those approved to the British specification D.Eng.R.D.2487 and the American specification MIL-L-7808, although the latter will in future be increasingly manufactured from polyol esters. The viscosity/temperature relationships of D.Eng.R.D.2487 and MIL-L-7808 oils are shown in Fig. 27.

Typical viscosity indices for di-ester base oils are between 150 and 165, but D.Eng.R.D.2487 is formulated with a polymeric or complex ester viscosity-index improver which raises the VI to about 175.

The volatility of di-esters varies considerably with their composition, and the sebacates and azelates have much lower volatility than adipates and glutarates of similar molecular weight. It can be assumed that the blends used in the standard jet engine lubricants are selected to give acceptably low volatility, since their use at high altitude and at high engine temperatures makes low volatility essential.

The fire resistance of di-esters is better than that of mineral oils, and comparable with that of the synthetic hydrocarbons, but they must still be recognized as flammable liquids. Typical flash points are 225°C or higher, and spontaneous ignition temperatures between 380 and 410°C.

The di-esters are generally better lubricants than mineral oils of similar viscosity, but their anti-wear and load-carrying properties are usually improved in fully-formulated jet engine lubricants by the use of tri-cresyl phosphate or similar additives.

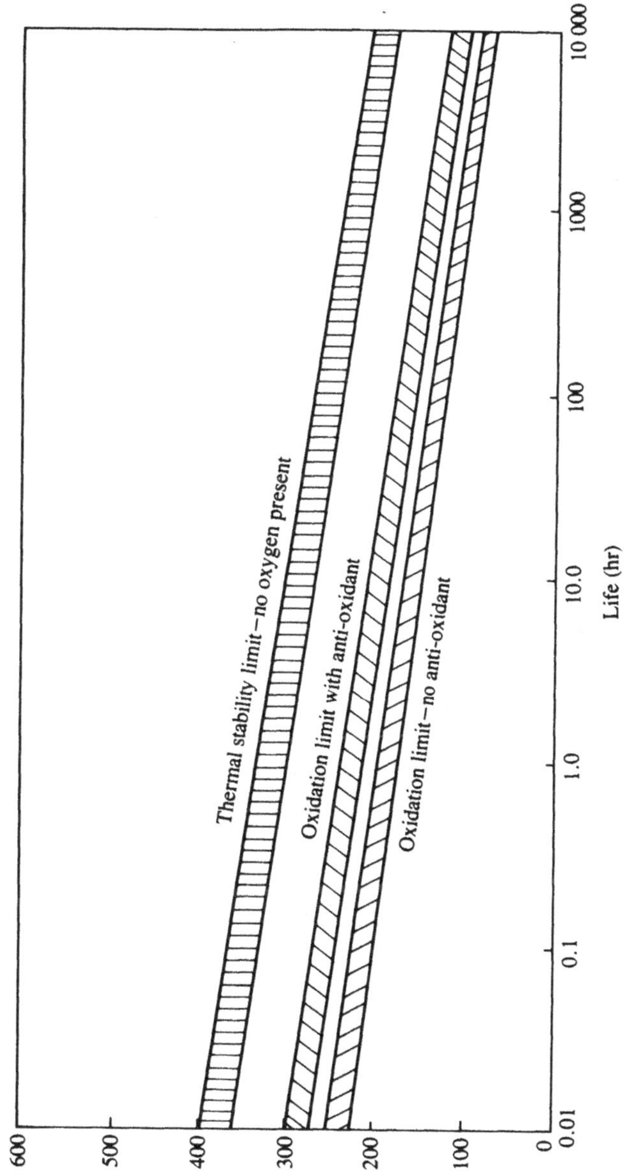

Fig. 26 Temperature limits for di-esters

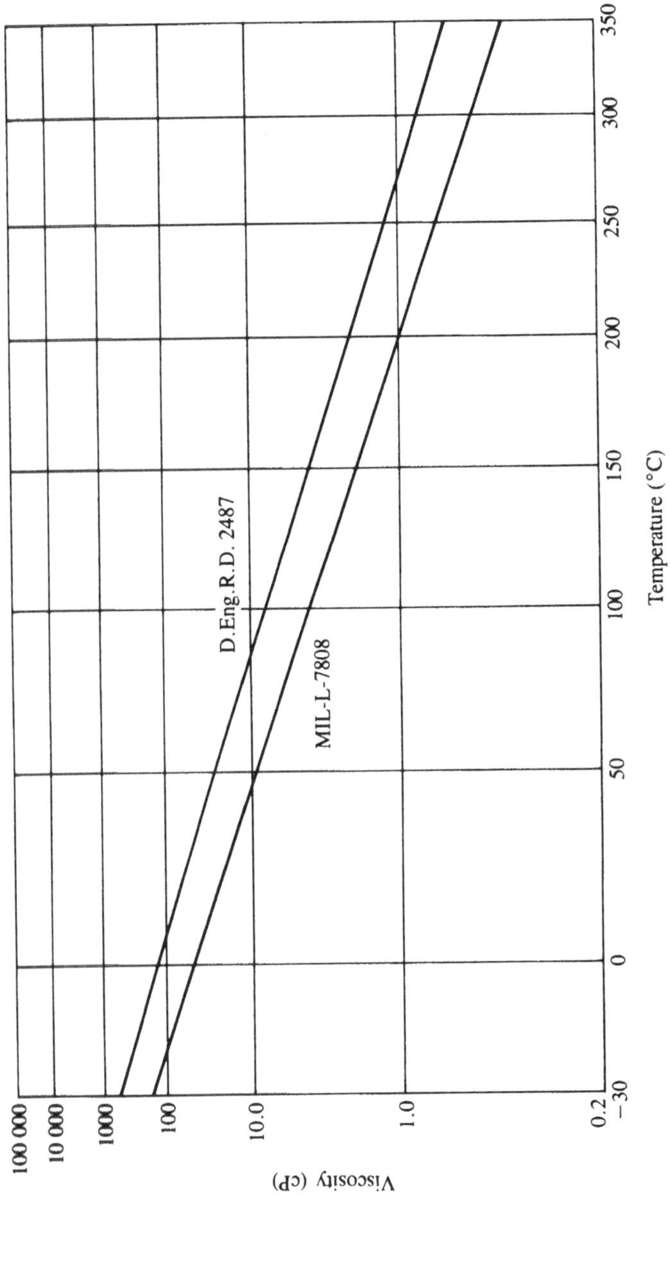

Fig. 27 Variation of viscosity with temperature for di-esters

7.5 POLYOL ESTERS

Typical polyol ester structures are shown in Fig. 25, and their thermal and oxidative degradation have been described in Chapters Four and Five. Temperature limits are in Fig. 28. There is some evidence that the effect of temperature change on thermal degradation rate is less than for the di-esters or mineral oils, and this is indicated by the flatter line for the thermal stability limit. However, the temperature limits can be seen to be higher than those for the di-esters and synthetic hydrocarbons.

The limits are again drawn broadly, because there is considerable variation between the different polyol esters, and especially between the pentaerythritol and trimethylolpropane esters. The initial temperatures for thermal degradation (T_D) have been reported as about 315°C, but the presence of certain metals reduces it to about 260°C.

The most readily available polyol esters are again those which are supplied as aircraft gas turbine lubricants, to the British specifications D.Eng.R.D.2497 and D.Eng.R.D.2499 and the American specification MIL-L-23699. These all have similar viscosities, and the variation of viscosity with temperature is shown in Fig. 29. There would be no difficulty in formulating polyol ester oils with higher viscosities if there were sufficient demand. The viscosity index is typically about 135–145.

The volatility of the polyol esters tends to be lower than that of the di-esters, in line with their general higher viscosities. Flash temperatures are between 230 and 280°C, and spontaneous ignition temperatures about 400°C. They are therefore more difficult to ignite than the mineral oils and synthetic hydrocarbons, but similar to the di-esters, and must be generally considered to be flammable.

The polyol esters are very good lubricants, the best being those with straight-chain alkyl groups attached to the acid groups.

7.6 POLYGLYCOLS

The structure of the polyglycols has been described in Chapter Four. They are basically of two types, the generally water-soluble ones whose molecules end in hydroxyl groups (diols), and those which terminate in alkyl groups and are therefore water-insoluble and partly oil-soluble (ethers).

The polyglycols do not have outstanding thermal and oxidative stab-

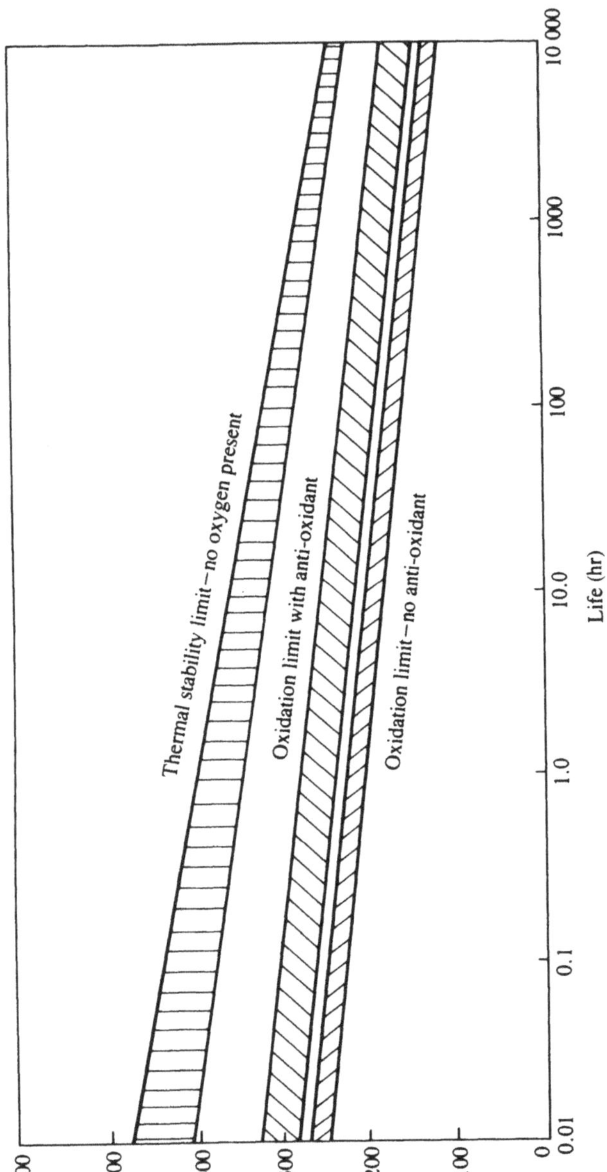

Fig. 28 Temperature limits for polyol esters

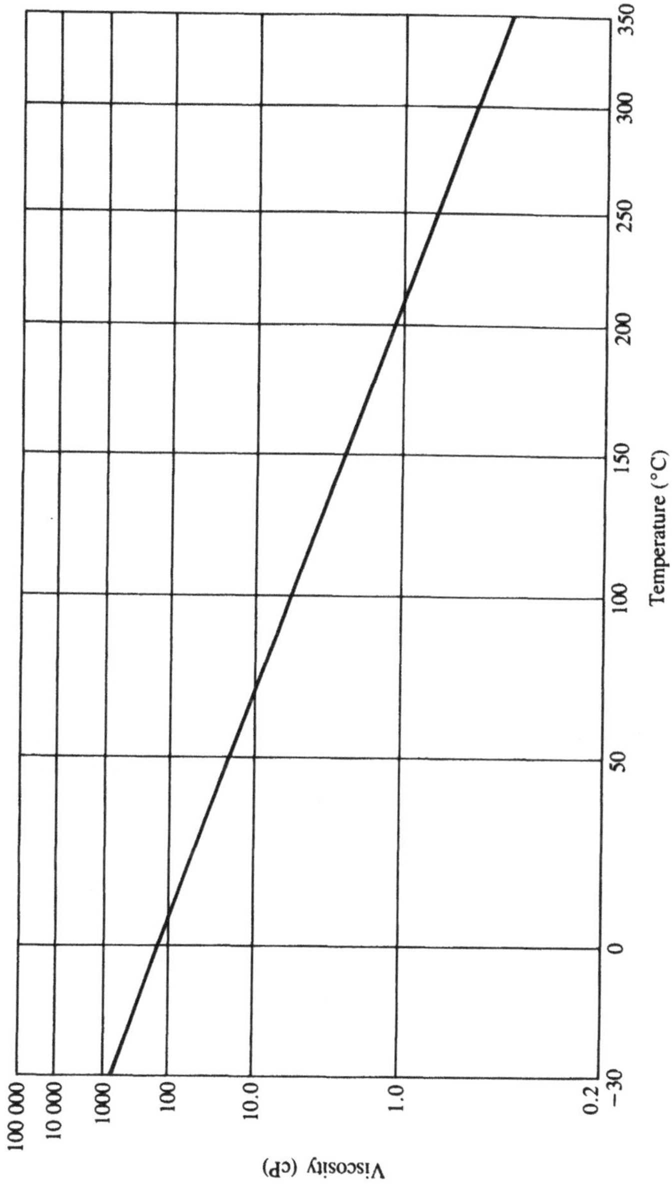

Fig. 29 Variation of viscosity with temperature for polyol ester (D.Eng.R.D.2497)

ility. The initial thermal decomposition temperature (T_D) is between about 220°C for the diols and about 260°C for the ethers. Approximate thermal and oxidative temperature limits are shown in Fig. 30, but these are only approximate levels, because there is not a lot of consistent information available. Decomposition leads to a loss of fluid, but usually little deterioration of the remaining fluid, so that the usable life is relatively long if the fluid level is maintained.

The great advantage of the polyglycols for high-temperature applications is that they decompose by an 'unzipping' mechanism which gives volatile low-molecular-weight decomposition products and leaves clean surfaces in the hot mechanisms. They are commonly used in oven conveyors, but probably more often as greases or pastes, which are described in the next chapter.

Some typical viscosity/temperature relationships are shown in Fig. 31. It can be seen that the graphs differ from most of the others in this chapter in being curved. For this reason the concept of viscosity index is not meaningful, and neither are the other simple viscosity/temperature descriptions such as ASTM slope and temperature coefficient of viscosity. Viscosity increases with the polymer chain length and it is possible to obtain polyglycols of widely differing viscosities. The curves in Fig. 31 are only examples of the wide range available. The higher viscosity diol polyglycols have high melting points, and form waxy solids at normal temperatures. They are often mixed with alkyl-terminated types to improve low-temperature performance.

The volatility of the polyglycols varies with molecular weight and, therefore, with viscosity, but is generally low. Evaporation at higher temperatures is mainly due to oxidation, and can be very much reduced by means of an effective anti-oxidant.

The flash points tend to be higher than for mineral oils of similar viscosity, but are strongly influenced by oxidation, which produces flammable volatile products. An uninhibited polyglycol may have a flash point in the range 200–260°C, but this will be raised by 25–70°C by the use of an effective anti-oxidant. Spontaneous ignition temperatures vary from 320°C to 400°C. The pure polyglycols must be considered to be flammable, but the diols are mixed with water to produce non-flammable hydraulic fluids and lubricants for ambient-temperature applications.

Polyglycols are quite good lubricants, comparable with mineral oils, but the additive technology is not as well developed, so that mineral gear oils with powerful additives have superior EP properties. However,

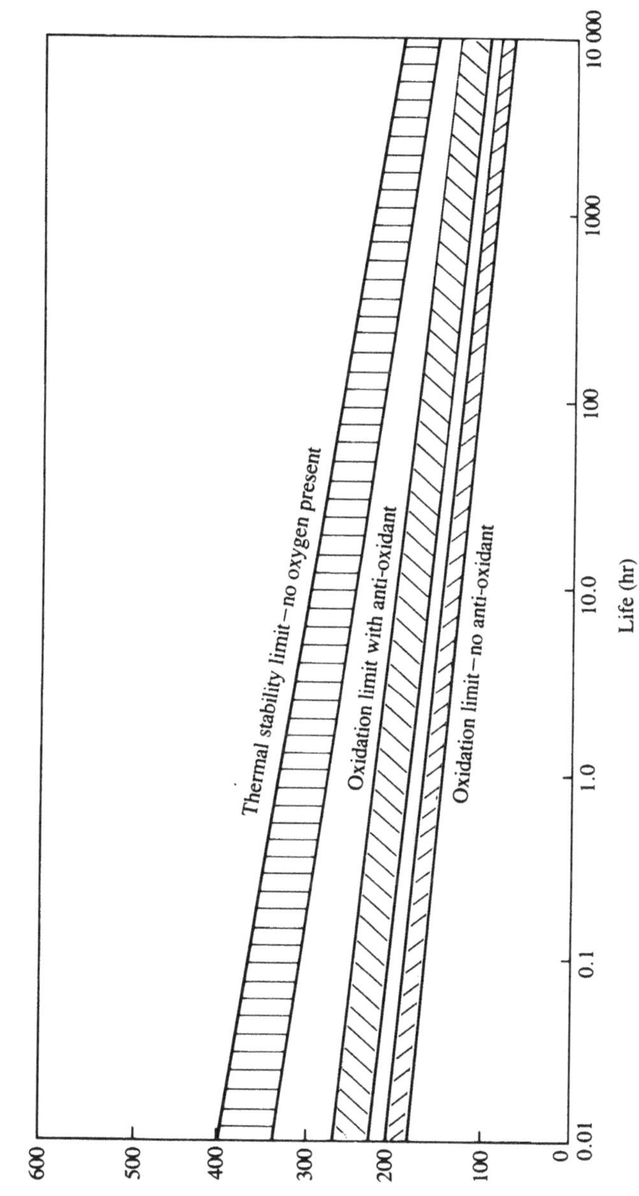

Fig. 30 Temperature limits for polyglycols

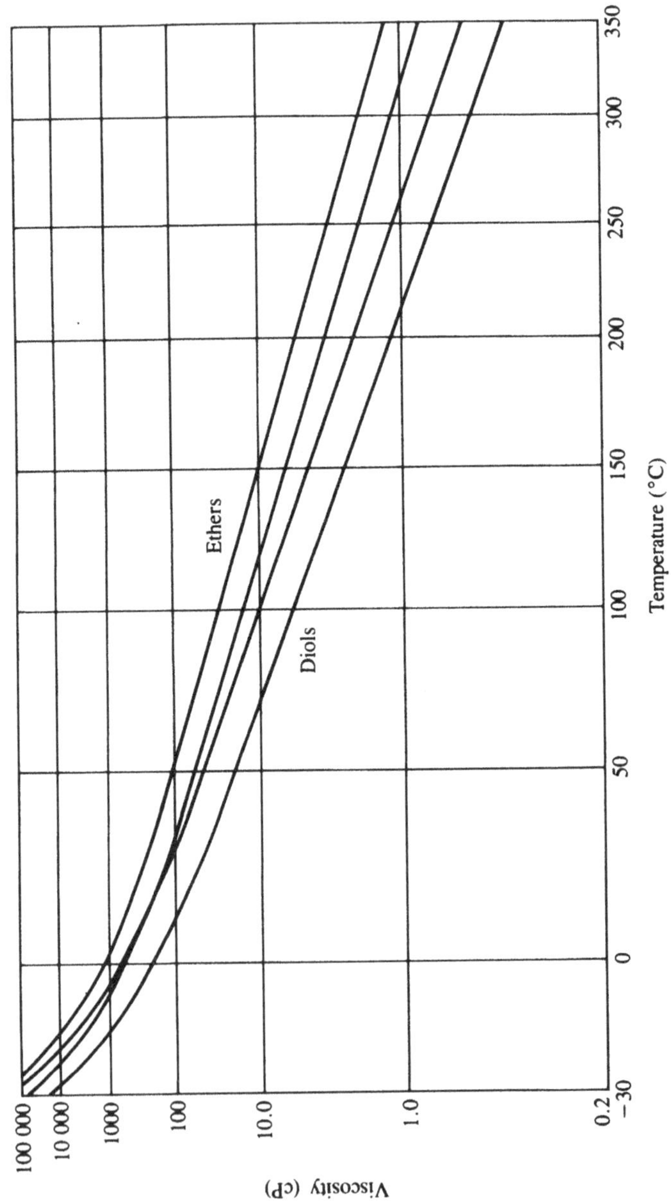

Fig. 31 Variation of viscosity with temperature for polyglycols

polypropylene glycols, usually ether-terminated, have the great advantage of giving very low friction in conditions of elastohydrodynamic, mixed, or boundary lubrication. They are therefore particularly useful for worm gears, where their low friction helps to reduce operating temperatures.

The hydroxy-terminated types (diols), especially the polyethylene glycols, have low solubility in petroleum liquids and other non-polar liquids. They are therefore very useful in pumps or compressors handling such materials, where they have much better resistance than mineral oils to washing out of the bearings or becoming diluted by dissolved gases.

7.7 PHOSPHATE ESTERS

The structure of phosphate esters has been explained in Chapter Four. Those which are used as anti-wear and EP additives are triaryl phosphates. These have better thermal and oxidative stability than the trialkyl phosphates, but poor viscosity indices. Phosphate ester lubricants and hydraulic fluids are usually based on mixed arylalkyl phosphates, sometimes with a viscosity index improver.

The high temperature properties of phosphate esters are generally inferior to the other lubricating oils described in this chapter. Their initial thermal decomposition temperatures (T_D) vary from 195°C for trialkyl phosphates to 420°C for triaryl phosphates, but these are considerably reduced in contact with many different metals, including steels. The thermal and oxidative stability temperature limits are very similar, and are plotted as a single band in Fig. 32. The levels shown assume contact with metals. Phosphate esters are included in this report, not because of any useful life at high temperatures, but because of their excellent fire-resistant properties, which are described later.

The viscosity/temperature characteristics of the pure phosphate esters vary enormously. The viscosity index of the triaryl phosphates tends to be low, increasing with the proportion of alkyl groups present, and increasing with the chain length of the straight-chain alkyl esters. The commercially-available products are usually mixed arylalkyl phosphates, and often contain a viscosity index improver. The resulting viscosity index can be as high as 230 for an aircraft hydraulic fluid, or as low as 0 for

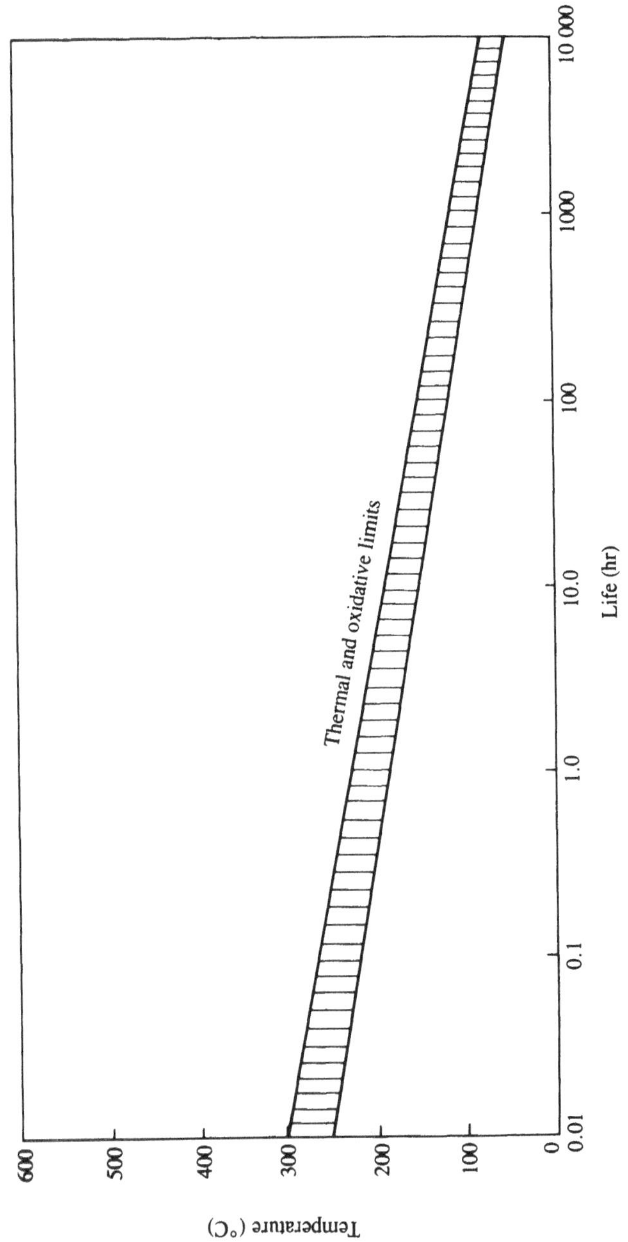

Fig. 32 Temperature limits for phosphate esters

an industrial fluid. Figure 33 shows some typical relationships for aircraft and industrial hydraulic fluids.

Phosphate esters are more volatile than mineral oils or esters having similar viscosities, but because they are not usable for any length of time at high temperature, evaporation is not normally a serious problem. They would, however, be unsuitable for use in hot vacuum applications.

Phosphate esters are highly fire-resistant in most circumstances, and their major use is in situations where there is a particular fire hazard, such as in aircraft hydraulic systems, industrial hydraulic systems in hot plant, and compressors.

They are excellent boundary lubricants, being of the same chemical type as some of the best anti-wear additives.

7.8 SILICONES

The structure of the silicones has been described in Chapter Four, consisting of a skeleton of straight chains of alternate silicon and oxygen atoms, to which different organic alkyl or aromatic groups are attached. The main types of organic side-group are methyl, phenyl, chlorinated phenyl, and trifluoropropyl, and the physical and chemical properties vary considerably with the nature of the side-group.

The dimethyl silicones may be considered as the fundamental class, and are the most widely used. Their initial thermal decomposition temperature T_D is about 315°C. The thermal stability is improved progressively by replacement of methyl side-groups by phenyl. Low and medium phenylmethyl silicones are not much better than the dimethyl silicones, but versions with more phenyl than methyl can have a T_D over 370°C.

Dimethyl and trifluoropropyl silicones oxidize slowly but significantly at 205°C. The oxidation resistance improves steadily as the methyl groups are replaced by phenyl or chlorophenyl, and the resulting products have good oxidation resistance to 260°C. The limits for thermal and oxidative stability in Fig. 34 are therefore shown separately for various classes. Some iron salts and ferrocene are effective anti-oxidants for silicones, but reduce the thermal stability, and are not widely used.

The initial thermal or oxidative degradation product of all the silicones

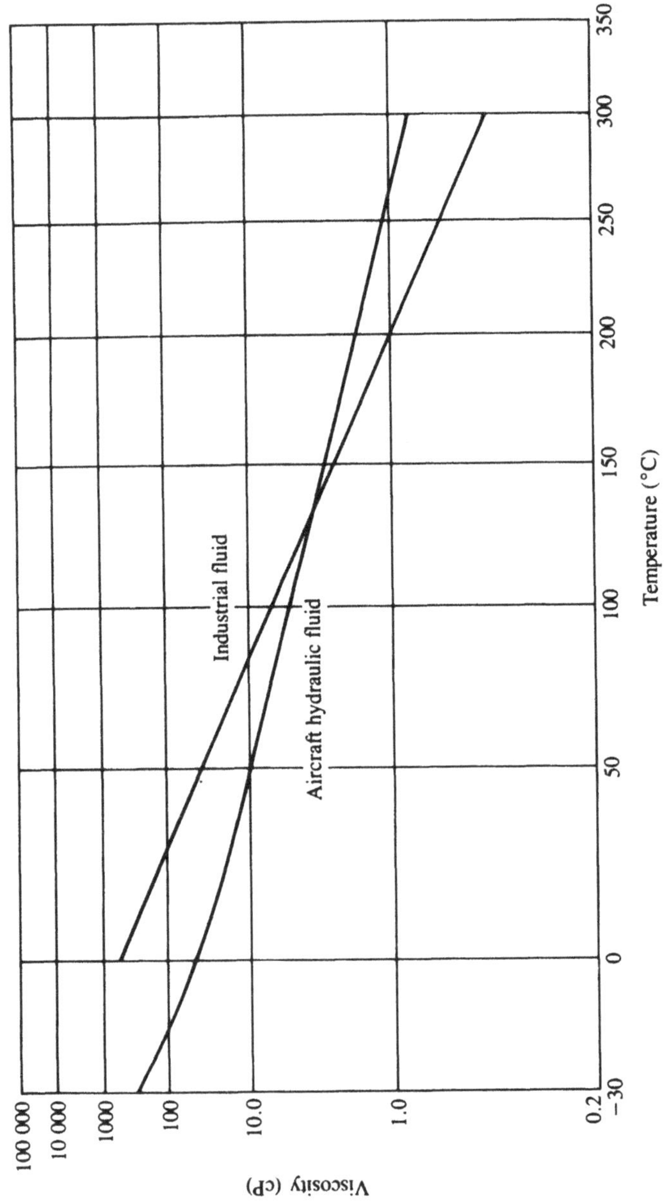

Fig. 33 Variation of viscosity with temperature for phosphate esters

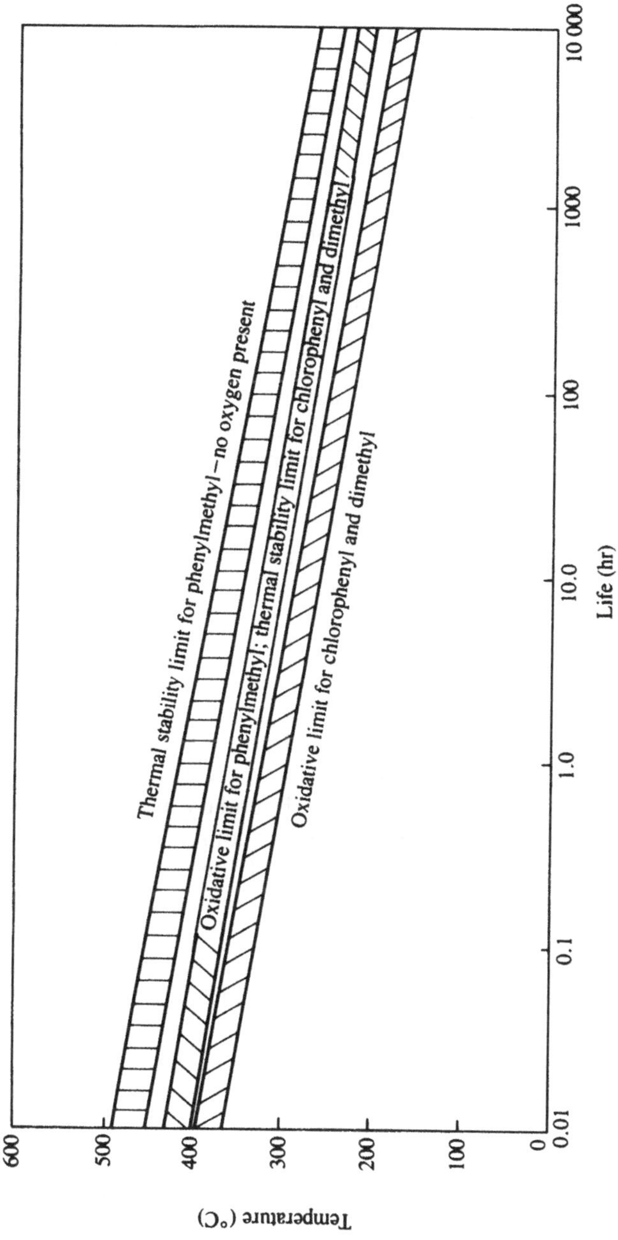

Fig. 34 Temperature limits for silicones

appears to be a gel-like cross-linked polymer. Under very mild sliding conditions the production of a lacquer on a surface can be achieved which improves operation under mild boundary conditions.

Under more severe rubbing conditions the cross-linked polymer may be formed with the phenylated silicones. With the dimethyl silicones, under these severe sliding conditions, the only degradation product found in a series of different tests was silica (silicon dioxide).

The viscosity–temperature characteristics also vary with the nature of the organic groups. Dimethyl silicones have the flattest viscosity–temperature curves of all oils except the perfluoropolyalkyl ethers, with VI over 260. They are also available in a wide range of viscosities from less than 1 cSt to 1 000 000 cSt at 25°C.

The phenylmethyl and chlorophenylmethyl silicones have inferior but still very good viscosity–temperature characteristics, with VI between 170 and 195, but they are available in only a few viscosity grades.

The volatility of all the silicones is low by comparison with mineral oils of similar viscosity. Even at the highest usable temperatures evaporation is not a problem, except for the lowest viscosity fluids, which are not viscous enough to be used as lubricants.

The flash point also varies considerably with viscosity, but is generally higher than for comparable mineral oils. The spontaneous ignition temperature of dimethyl silicones is about 480°C, depending on the type of test method. Once ignited all the silicones except the chlorophenyl and trifluoropropyl forms will continue to burn.

Silicones are satisfactory lubricants where the lubrication is hydrodynamic or elastohydrodynamic. For boundary lubrication it is safest to regard them as poor. There are metal combinations for which good results have been reported, but most of these combinations are ones which are not normally used for bearings or gears, especially at high temperatures.

The dimethyl silicones are the worst boundary lubricants, especially for steel against steel, and even at low loads and speeds they lead to unpredictable high friction and seizure. If very carefully run in, it is possible to generate a cross-linked silicone lacquer on bearing surfaces, and this enables the bearings to run successfully under mild conditions.

Phenylmethyl silicones have been reported to be better boundary lubricants than the dimethyl silicones. The greatest improvement is obtained with the chlorophenyl and trifluoropropyl silicones. These have useful anti-wear properties, and even EP properties at high chlorine or fluorine levels, but in the presence of even small amounts of moisture the chlorophenyl forms can cause corrosion.

None of the usual anti-wear or EP additives is effective in silicones. Some organometallic compounds have been found to improve boundary lubrication, but with a reduction in thermal and oxidative stability.

7.9 SILICATE ESTERS AND DISILOXANES

The silicate esters are esters of orthosilicic acid, and the disiloxanes are esters of the dimeric form of orthosilicic acid.

Tetra-aryl silicates are used as heat-exchange fluids because of their high boiling points and thermal stability. The initial thermal decomposition temperature T_D has been quoted as high as 450°C. They are not used as lubricants because they are solid at normal temperatures. There seems to be no reason why they could not be used in sealed high-temperature systems where no low-temperature use is necessary, but they are not suitable for systems which are not sealed or inerted, because oxidation resistance is not outstanding.

The alkyl silicates have poorer thermal stability, with T_D of the order of 345°C, but they have much wider liquid ranges and have been used as hydraulic fluids for a few high-speed aircraft. The temperature limits shown in Fig. 20 for mineral oils probably apply approximately to the silicate esters.

The viscosity/temperature characteristics of alkyl silicate esters are quite good, with viscosity indices in the range 170–230. Figure 35 shows the viscosity–temperature relationship for an alkyl silicate aircraft hydraulic fluid.

The volatility is slightly higher than for equivalent mineral oils, and in most respects the flammability characteristics are comparable with those of mineral oils.

The lubricating characteristics are comparable with those of paraffinic mineral oils without additives, in other words, they are only fair. The solubility of additives in the alkyl silicates is good, but there seems to be little published information about the benefits of anti-wear or EP additives. There is no evidence that the fully-formulated fluids can compare in performance with mineral oils with appropriate additives.

A major disadvantage of the silicate esters is their poor hydrolytic

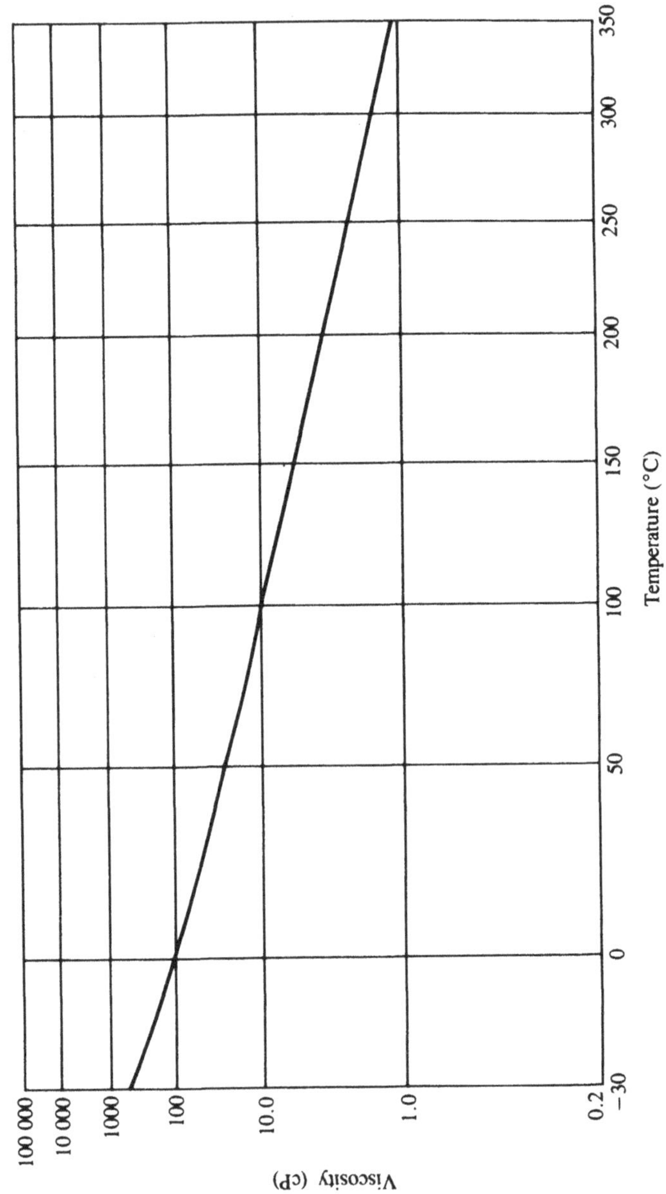

Fig. 35 Variation of viscosity with temperature for silicate ester

stability. This appears to be associated with oxidation, and the use of certain anti-oxidants improves the hydrolytic stability.

7.10 POLYPHENYL ETHERS

The structure of polyphenyl ethers has been described in Chapter Four. Their initial thermal decomposition temperatures, T_D, are in the range 440–485°C and the m5P4E type used for lubrication has T_D of 470°C. Oxidation resistance is also very high, and approximate temperature limits are shown in Fig. 36. The search for antioxidants has not been very successful, but tetraphenyl tin has been reported to be effective, and the presence of some metals reduces oxidation.

The viscosity/temperature characteristics are poor, with VI about −40. Their main disadvantage in this respect is that in order to achieve useful viscosity at their highest operating temperatures it is necessary to use the larger molecules, which are solid at normal temperatures. Individual 5P4E and 4P3E polyphenyl ethers have melting points between 40 and 150°C. Mixtures can be produced which are liquid at normal temperatures, but with pour points of 5°C. The one type which has been used successfully as an aircraft hydraulic fluid has had to be heated above its pour point before use. Figure 37 shows the viscosity–temperature relationships for two potential lubricant-grade polyphenyl ethers.

The volatility is also dependent on molecular size. At comparable viscosities the polyphenyl ethers are all less volatile than the various esters, but at the high temperatures at which they are capable of operating, significant evaporation loss is possible. In the ASTM D972 test for evaporation, the m5P4E polyphenyl ether had the same evaporative loss at 285°C as a low-viscosity di-ester has at 205°C.

A wide variety of viscosity grades is theoretically available by blending the individual molecular types, but the availability of polyphenyl ethers at present is limited, and the cost is correspondingly high.

The polyphenyl ethers are generally less flammable than the mineral oils and esters. Flash points are 285°C for m5P4E and 230°C for m4P3E, and spontaneous ignition temperatures are 570°C and 590°C, respectively. The flash points are well below their potential operating temperatures, so that it is important to avoid contact between the oil and a naked flame at those temperatures.

Boundary lubrication with the polyphenyl ethers is only fair, far better

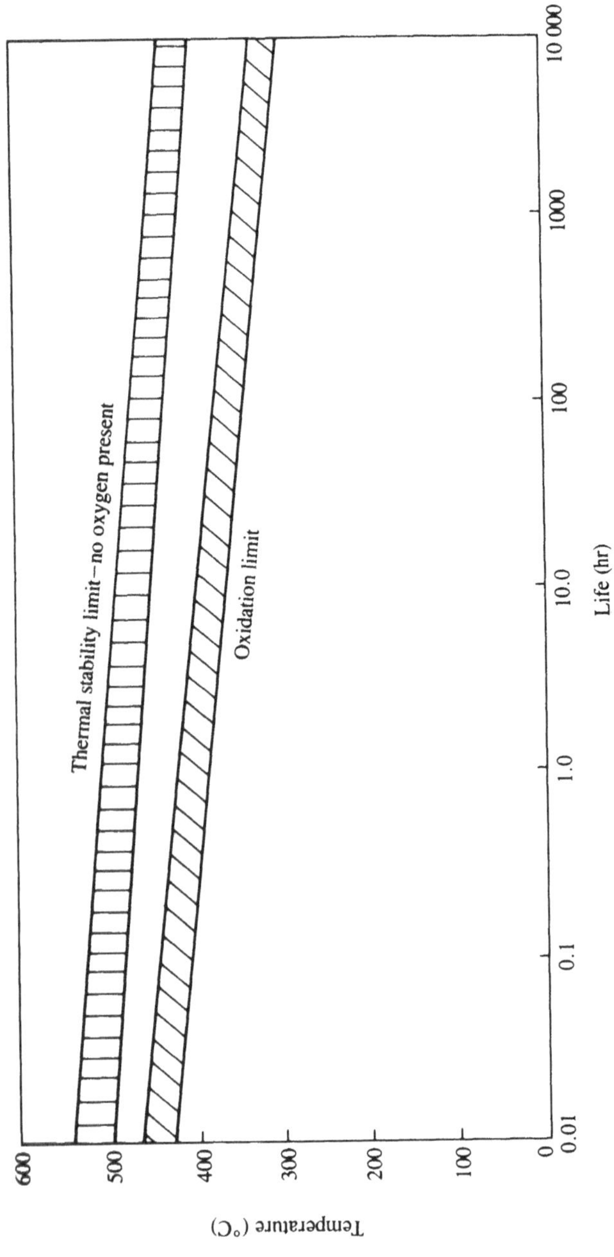

Fig. 36 Temperature limits for polyphenyl ethers

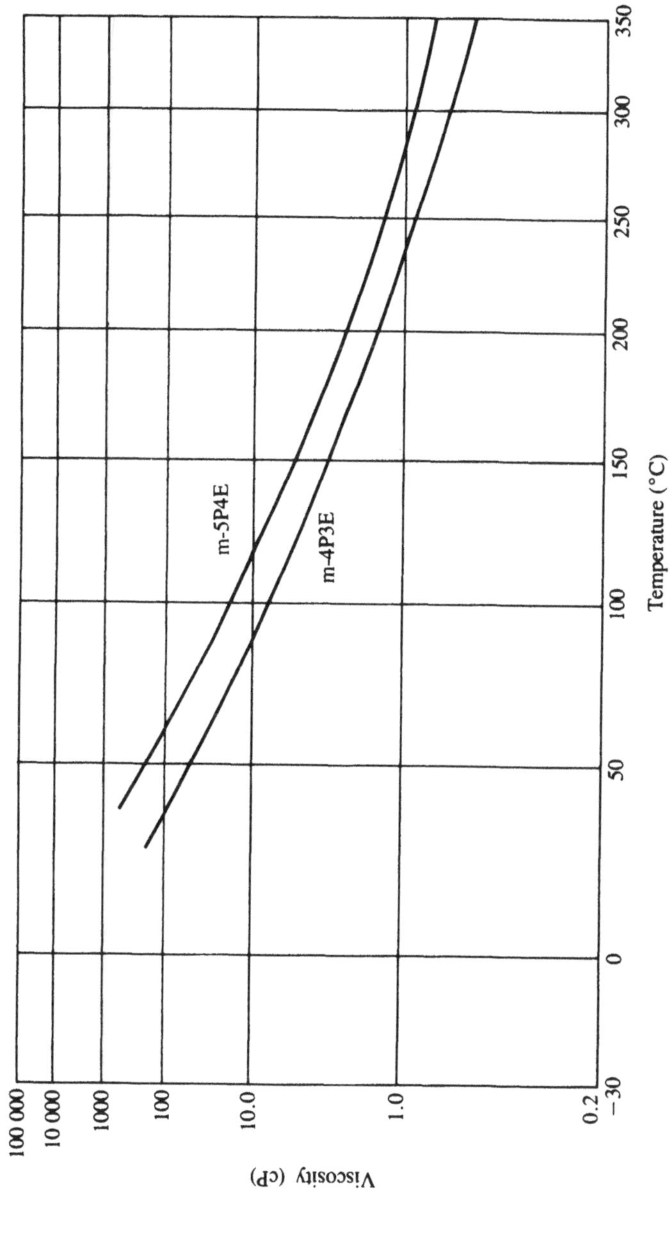

Fig. 37 Variation of viscosity with temperature for polyphenyl ethers

than the silicones, but slightly inferior to di-esters and paraffinic mineral oils without additives. The search for useful anti-wear and EP additives has not been very successful, so that the performance of fully-formulated polyphenyl ethers is very much inferior to that of fully-formulated mineral oils or esters. In addition, there is a tendency to produce a sludge when sliding takes place at high temperature in the presence of oxygen.

At the highest temperatures at which the polyphenyl ethers can be used there are no better boundary lubricants available, but it is important to ensure, wherever possible, that bearings and gears are operating in hydrodynamic or elastohydrodynamic lubrication, where only the viscosity characteristics are important.

7.11 C-ETHERS

C-Ethers are compounds similar in structure to the polyphenyl ethers, but in which some or all of the oxygen ether linkages are replaced by sulphur atoms. This gives lower pour points and better boundary lubrication, but at the expense of reduced thermal and oxidative stability.

The initial thermal decomposition temperature, T_D, of the C-ethers is about 390°C, but the oxidation resistance seems to be only slightly lower than that of the polyphenyl ethers.

Insufficient information has been obtained to enable temperature limit curves or viscosity–temperature curves to be provided, but it seems probable that the oxidation limit will be similar to that of the polyphenyl ethers in Fig. 36.

7.12 PERFLUOROPOLYALKYL ETHERS (PFPAE)

The perfluoropolyalkyl ethers (PFPAE) are also variously known as perfluoroalkyl ethers (PFAE) or perfluoroalkyl polyethers or perfluoropolyethers (PFPE).

Their general structure has been described in Chapter Four, but there are several different types which differ in their chemical stability and viscosity–temperature characteristics. The three main categories are, first, the Krytox type, manufactured by DuPont, and Fomblin Y, manufactured by Montedison (Ausimont), which have a repeating isopropyl (C_3) group with side-chain methyl groups; second, the Fomblin Z type from Montedison (Ausimont), which has repeating ethyl (C_2) and methyl (C_1) groups with no methyl side-groups; and thirdly, the Demnum type from Daikin with n-propyl (C_3) groups and no methyl side-groups. There have also been other types, which have been synthesized in small quantities for research studies.

The PFAE fluids have been available for over thirty years, but their extremely high cost has limited the range of applications to those such as oxygen systems where the quantities required are very small. They appear to have the best potential of all existing fluids for high temperature lubrication, although there are serious problems to be overcome, and great efforts are at present being made to develop lower-cost manufacturing techniques.

Initial thermal degradation temperatures, T_D, have been reported as 390°C for the Krytox type and 360°C for the Fomblin Z type. Oxidation takes place at lower temperatures, as shown in Fig. 38. Thermal degradation is not affected by the presence of metals, but oxidation is very much increased by metals, including steel and titanium. Two antioxidants have been found to be very effective at 288°C, but relatively ineffective at 316°C. These are a perfluorophenyl phosphine and a monophosphor-s-triazine, each linked to perfluoropolyalkyl ether chains to improve their solubility in the PFAE fluids.

Like the polyglycols, the PFAE fluids decompose to give mainly volatile low-molecular-weight degradation products which are lost from the system. The result of degradation is, therefore, fluid loss, with only a very slight deterioration in chemical and viscosity properties. It follows that the limits in Fig. 38 should be seen as indicating unacceptable rates of fluid loss (about 15 per cent) rather than deterioration in properties.

The viscosity of all three types varies with chain length (i.e., molecular weight), and the size of the repeating unit (C_1, C_2, or C_3), but only slightly with the presence or absence of branching. The viscosity index for various straight-chain Fomblin fluids varies between 286 and 360, and for some of the Krytox and Demnum fluids appears to vary between about 70 and 140. Typical viscosity–temperature relationships are shown in Fig. 39. It should be remembered that a wide range of molecular weights is available in all three types, and a viscosity grade can be chosen to give either the best high-temperature viscosity or the best low-temperature flow. However, this choice is less critical for the straight-chain Fomblin fluids because of their high viscosity indices.

The volatility also varies with structure and molecular weight. Because

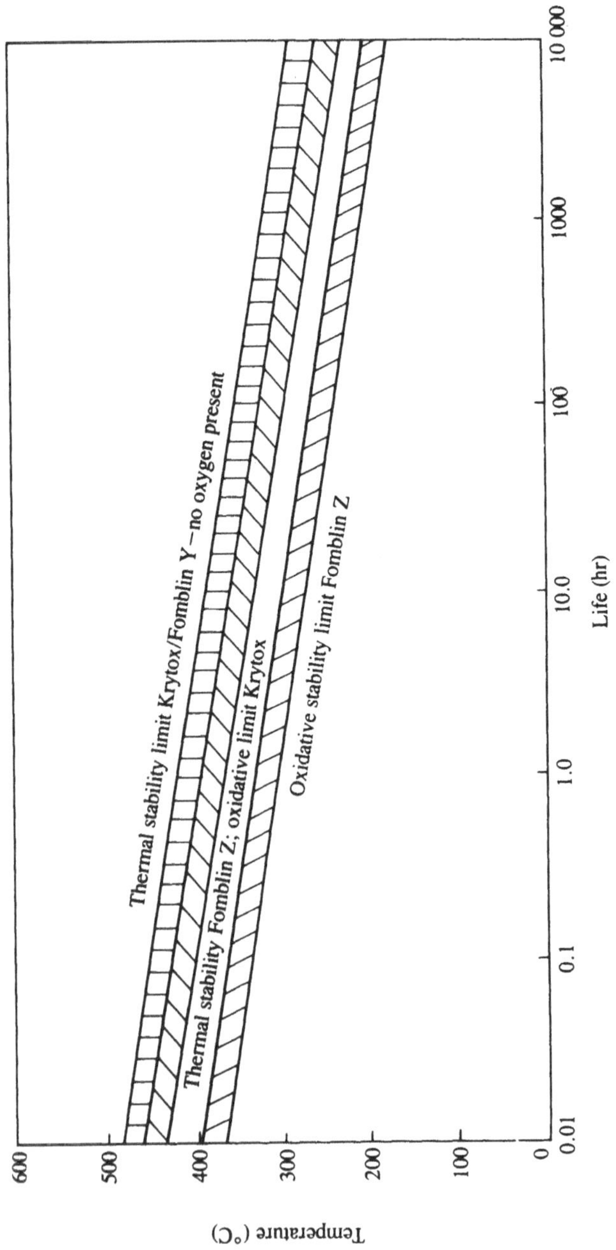

Fig. 38 Temperature limits for perfluoropolyalkyl ethers

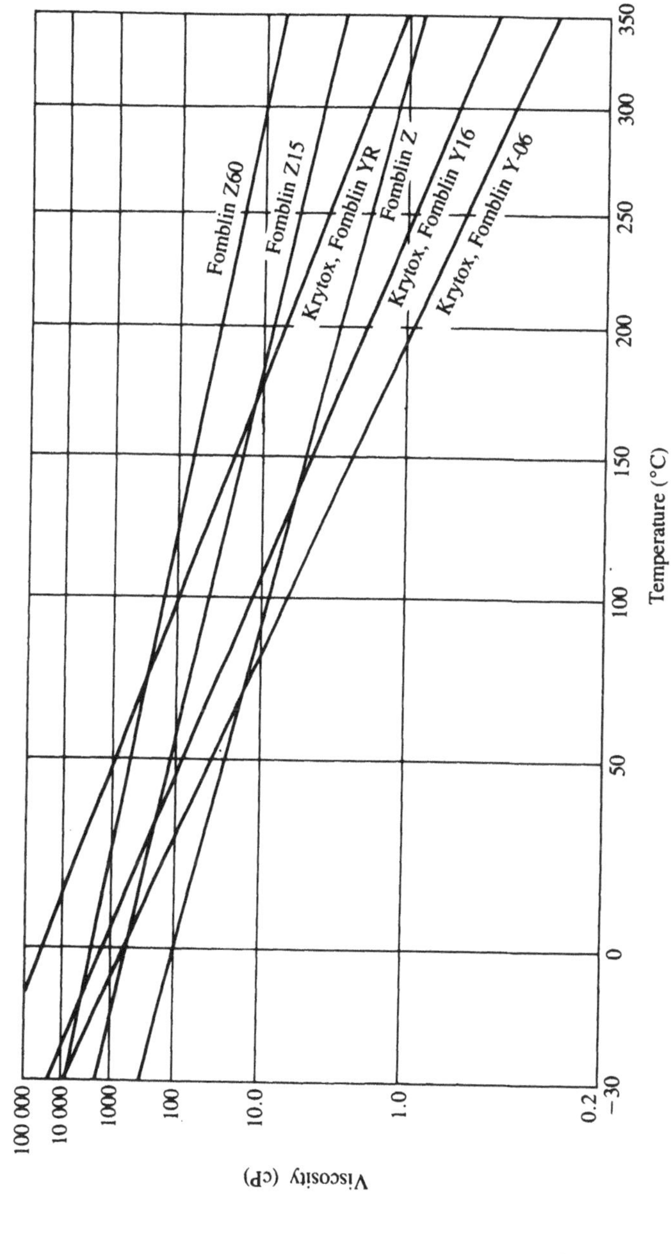

Fig. 39 Variation of viscosity with temperature for perfluoropolyalkyl ethers

of lack of data it is difficult to make detailed comparisons, but the limited data available suggests that for similar molecular weights the branched-chain fluids have lower volatility. However, if a selection is made on a viscosity range basis, the far better viscosity–temperature relationship of the straight-chain Fomblin fluids makes it possible to use a much higher molecular weight and, therefore, a lower volatility. For a Fomblin fluid with an operating temperature range from −54°C to 370°C the evaporation in a stream of air at 204°C was only about 1.2 per cent in 22 hours.

The flammability of the PFAE fluids is very low, with flash points over 350°C and spontaneous ignition temperatures higher than 700°C. For practical purposes they can be considered virtually non-flammable.

The boundary lubrication characteristics are very poor, and seizure can take place even at low loads and speeds. At high temperatures the fluids have been shown to remove surface oxide from various metals, and under sliding conditions this would obviously lead to rapid seizure. Under hydrodynamic conditions satisfactory lubrication should depend only on the choice of an adequate viscosity grade, as with all other fluids. However, there have been suspicions that even under conditions which should give full but thin fluid film separation, some unpredictable seizures have occurred.

Existing conventional anti-wear and EP additives are not soluble in the PFAE fluids. One successful anti-wear additive has been reported, but its chemical composition was not stated, and since the wear test was only at 75°C it was not very relevant for a high-temperature lubricant.

7.13 OTHER ORGANIC LIQUIDS

There are a number of other organic liquids which have been used or studied as lubricants, but none appears to have any great promise for use at high temperatures. Some of the more important ones are listed in Table 6.

7.14 LIQUID GLASSES

The limited temperature range for which any particular lubricant is liquid has been mentioned previously, especially for the case of polyphenyl

Table 6 Some additional organic liquids

Oil	*Remarks*	*Application*
Rapeseed oil	Low oxidation temperature, good boundary lubrication, non-carburizing	Hot metal-working
Castor oil	Low oxidation temperature, good boundary lubrication	Racing engines
Chlorinated diphenyls	Fire resistant, fair boundary lubrication	Hydraulic fluids, compressors
Chlorofluorocarbons	Fire resistant, T_D 246°C	Experimental hydraulic fluid
Alkyl silanes/ silahydrocarbons	Good lubricants, mediocre oxidation resistance	Experimental hydraulic fluid
Fluoroesters	High thermal and oxidation stability, poor viscosity–temperature properties	Experimental engine lubricants

ethers. For temperatures above 500°C it seems unlikely that any oil will ever be developed which is also liquid at normal temperatures.

Once this situation is accepted, it is possible to extend the search for high-temperature lubricating liquids to a wide variety of inorganic materials which are solid at lower temperatures. The main criteria for selection then become the viscosity at the required temperature and suitable chemical stability. Three groups of materials which have been found suitable in these and other respects are glasses, liquid metals, and certain oxides.

In Chapter Two it was explained that amorphous solids do not have a clear melting point, but soften progressively over a wide temperature range, thus giving a very wide range of viscosities. Glasses are an extreme example of this phenomenon. As a molten glass cools, its viscosity increases progressively without any discontinuity, until its character is clearly solid.

Some workers in fact consider that a glass is not a true solid, but a supercooled liquid of very high viscosity. One factor which supports this argument is that if a glass is worked extensively while in a soft condition it can crystallize, and then ceases to be glassy. Other authorities consider that a glass is an extreme example of an amorphous solid.

Figure 40 shows the variation of viscosity with temperature of several commercial glasses. It can be seen that the viscosities vary continuously over a range from 100 P to 10^{15} P. Obviously, if a glass is to be used as a

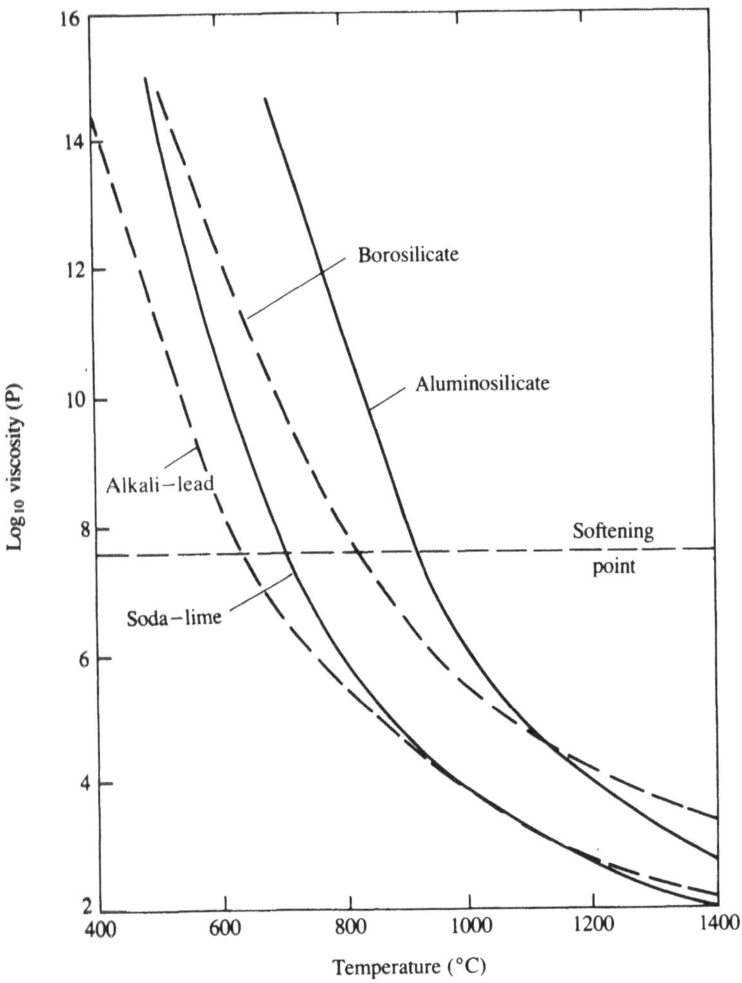

Fig. 40 Viscosity of some commercial silicate glasses *(From R. H. Doremus,* Glass Science, *1973. Reproduced by permission of John Wiley and Sons)*

liquid lubricant, it is the low-viscosity part of this range which is important, and for all the glasses in Fig. 40 viscosities of 1000 P or lower only occur at temperatures higher than 1000°C.

It follows that the main use of liquid glass for lubrication has been in the hot working of steels and other high-melting metals at temperatures over 1000°C.

All normal glasses are made from fully-oxidized materials, such as silica, boric oxide, calcium oxide (lime), alumina, magnesia (magnesium oxide), and lead oxide. They are therefore not oxidizable, and there is no temperature at which they decompose oxidatively or thermally. They can, however, attack other materials, either chemically or by dissolving them.

The variety of glasses and their interactions is too great to be covered in detail. Obviously molten glasses cannot be used to lubricate ceramics such as silica and alumina which are themselves glass constituents. They will also attack some of the more reactive non-ferrous metals, such as copper.

It seems to be necessary for a molten glass to wet a surface to some extent in order to lubricate it. The result is that when the metal stock or mould cools, there will be an adherent film of glass solidified onto it. This can usually be broken away fairly cleanly by impact or by rolling, but the process is improved if the glass is chosen to have a coefficient of thermal expansion as different as possible from the materials being worked or the materials of the moulds.

The handling of the glass in solid form is more difficult than for a liquid, but pre-melting of the glass is generally considered uneconomical. The heated metal stock may be allowed to slide over powdered or granular glass which is picked up on the softened metal surface. Alternatively a mat of glass fibre may be positioned at the entry of an extrusion mould.

7.15 LIQUEFIED OXIDES

It is theoretically possible to use as a liquid lubricant any material which is liquid and chemically stable in the right temperature range. In practice most of the useful ones at high temperatures are oxides because other classes of compound tend not be fully stable in contact with air, or to react with possible bearing materials. These useful oxides include some of the 'salts' which are effectively compound oxides, such as sodium metasilicate ($Na_2O{:}SiO_2$). A few others are fluorides.

Table 7 lists some of the oxides and salts which are potentially usable as liquid lubricants, and Fig. 41 shows the variation of viscosity with

Table 7 Some liquefiable oxides and salts

Oxide or salt	*Melting point* (°C)
Boric oxide B_2O_3	450
Sodium carbonate Na_2CO_3	851
Potassium fluoride KF	875
Lead oxide PbO	890
Lithium fluoride	896
Sodium metaborate	966
Potassium metasilicate	976
Sodium metaphosphate	988
Sodium fluoride NaF	992
Tin oxide	1042
Sodium metasilicate	1087
Germanium oxide	*ca* 1100

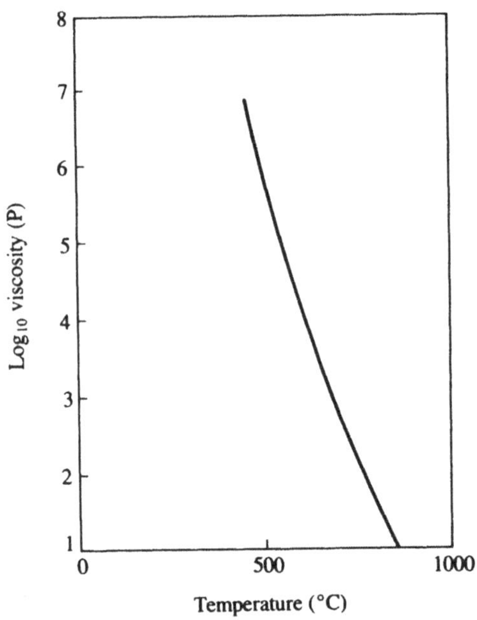

Fig. 41 Viscosity of molten boric oxide

Table 8 Some possible liquid metal lubricants

Metal	*Melting point (°C)*	*Boiling point (°C)*
Mercury	−39	357
Rubidium	39	686
Potassium	63	760
Sodium	98	883
Indium	157	2080
Lithium	181	1342
Tin	232	2270
Lead	328	1740
Zinc	420	907
Aluminium	660	2467
Silver	962	2212
Gold	1064	3080
Manganese	1244	1962

temperature for the most useful of them, boric oxide. However, even for boric oxide it can be seen that the viscosity only falls below 1000 cP at temperatures greater than 860°C, and below 100 cP above 1000°C.

7.16 LIQUID METALS

The use of liquefied metals as lubricants has been very limited. The range of metals available, and the resulting range of temperatures, are very high, and Table 8 lists thirteen of the most promising, with their melting and boiling points.

However, most metals oxidize quite rapidly in the liquid state, so that the choice for use in contact with air is limited to a very few, such as gold, indium, lead, and mercury. Even in the absence of air problems can arise due to chemical or dissolving attack on bearing materials and other surfaces.

Probably the greatest effort has gone into the use of liquid sodium and of a sodium–potassium alloy (NaK) containing 78 per cent of sodium. This was aimed principally towards use in nuclear reactors. Overall these two liquids showed satisfactory compatibility with a wider range of possible bearing metals than some of the other liquid metals such as mercury, rubidium, and lithium. They were also much more readily available and therefore cheaper.

The variation of viscosity with temperature for several liquid metals is

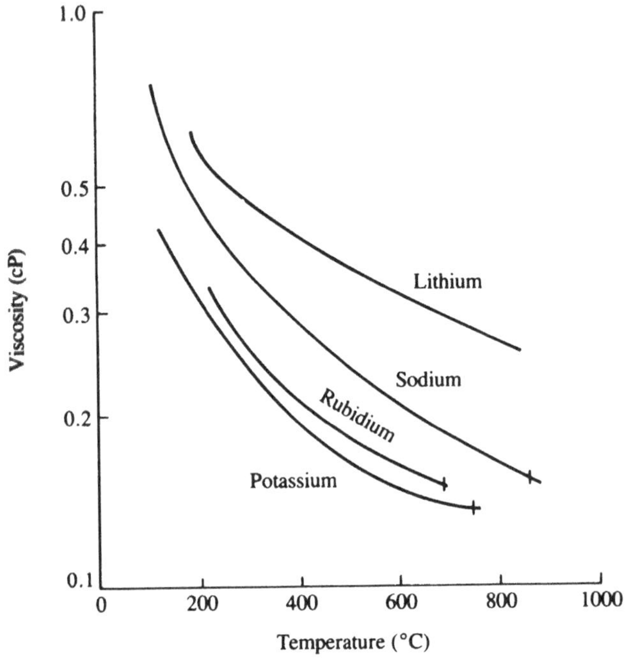

Fig. 42 Viscosity of some liquid metals

shown in Fig. 42. Boundary lubrication is generally not good, with coefficients of friction in the range 0.2–0.7 for a variety of bearing materials, and a tendency to adhesive wear. There are indications of some useful boundary lubrication with sodium containing a small concentration of oxygen, and this may involve the formation of a surface film of a double oxide.

CHAPTER EIGHT

Greases and Pastes

8.1 GENERAL

A lubricating grease is basically a liquid lubricant which has been thickened to a semi-solid consistency by means of a thickener. The high temperature behaviour of greases can, therefore, conveniently be considered in three parts. These are the aspects determined by the behaviour of the liquid, those determined by the properties of the thickener, and those which are related to the grease as a whole and not to either of the major components individually.

All the liquids which may be used as base oils for greases have been discussed in detail in Chapter Seven, and little more needs to be said about them here. The high temperature performance of greases is rarely, if ever, limited by the behaviour of the thickener itself. The gel structure, which is a function of base oil, thickener, and often other materials, can impose important limitations on high temperature performance of a grease, and this aspect will be described later.

Apart from the structure of a grease, its consistency has a direct and important influence on its high temperature behaviour, because it virtually prevents flow. The consistency of a grease, which is described in terms of penetration, can vary from very, very soft, penetration greater than 430, to very hard, penetration less than 160.

The softer greases, with penetrations greater than about 250, can be made to flow through pipes under a high enough pressure differential, and greases in large enough quantities will flow, or 'slump', under gravity. However, even the softest will not flow through confined spaces except under pressure. In particular, no convective flow is possible.

It follows that a major potential contribution to the cooling of bearings or other lubricated systems is not available when greases are used. Cooling is only by conduction or radiation, and that means that initially the removal of heat from the lubricated zone is mainly by conduction, either through the components or the grease itself.

Most greases are, in fact, very poor conductors, since the conductivity is determined primarily by the continuous phase, which is the base oil. Where thermal conductivity is particularly important, additives such as graphite or copper powder may be used to improve the conductivity. Such additives are not used in the softest greases because it is difficult to keep them in stable dispersion. In the harder greases they have some adverse effect on flow properties, so that they are not suitable for use in high-speed bearings.

Obviously, in any situation where the effects of high temperature are likely to impose limitations, it is important to minimize all types of heat input. It will therefore be instructive to consider the four categories of heat input discussed previously in Chapter Two in the specific context of grease lubrication. However, there are also two heat sources which are specific to grease lubrication.

Because of the high effective viscosity of a grease, any prolonged mechanical working leads to a high viscous power dissipation and, therefore, to considerable heat generation. As a result it is normal good practice in grease lubrication to avoid filling a bearing or gearbox with grease. The grease supplied should provide a thin film on the surfaces of the moving components, while the quantity remaining should be small enough to enable it to escape into unswept (i.e., undisturbed) spaces such as the covers of a rolling bearing or any free space in a gearbox.

Any excess of grease which cannot be accommodated in this way will be continually mechanically worked, for example, by the rolling elements of a bearing or the teeth of a gear. This working is known as 'churning' and can rapidly lead to overheating and failure of a grease, even in a low-temperature application.

The second problem is called 'slumping'. The ability of a grease to remain clear of the moving surfaces in a bearing or gearbox depends on the consistency and the volume present. A large quantity of soft grease will tend to fall, or slump, into the path of the moving components. It will probably then be thrown back into the free spaces, but a continued repetition of this process leads to a particular form of churning with resultant unwanted heating.

The problem of slumping is usually overcome by the use of a harder grease which will not slump in the volume present. Where the use of a harder grease is undesirable for other reasons, such as excessive starting torque, the problem can sometimes be solved by the direct application of a thin film of grease to the moving surfaces, without any surplus being present. This technique is sometimes called 'grease plating'.

The four categories of heat input discussed in Chapter Two all apply to grease lubrication. The effects of ambient temperature do not differ in any way from those in oil lubrication, but the effects of the other three are all different for greases.

Most grease-lubricated components have no facility for re-lubrication other than the use of a grease-gun feeding through a nipple. For such components there is no system heat input as previously described. A relatively few machines are equipped with pressure-feed systems in which the grease is re-supplied through pipes. Such systems are especially likely to be used for high temperature applications because of the difficulty of supplying such applications manually.

As far as possible the pipes carrying the grease feed should be routed to avoid hot areas. This will not always be possible, and the last part of the feed pipe, adjacent to a hot bearing or other components, will inevitably be subjected to a temperature approaching that of the component.

External insulation of the feed pipe is not very effective because of conduction of heat along the walls of the pipe and other structures, and because the residence time of the grease at that point is likely to be long enough for the temperature to reach equilibrium. The use of glass or ceramic pipes is more likely to be effective, provided that the risk of pipe fracture can be avoided. Otherwise, the only useful advice is to bear in mind the residence time of the grease in that hot location, and to realize that it may form a major part of the available life of the grease at the relevant temperature. In other words it must be recognized that the grease may have suffered significant degradation before it even reaches the bearing.

The effect of component heat inputs differs in two respects from their effect in an oil circulation system. The first is that all the heat input remains in the bearing with the grease, except to the very minor extent that the grease may be exuded through a seal or a grease valve.

The second is related to the peculiar nature of grease lubrication, in which the lubricant is present only as a thin film on the moving surfaces, and the only continuous medium between the moving surfaces and the housing is air or some other gas. Heat transfer from the lubricated surfaces is, therefore, only by conduction through the components or by radiation from the components to the walls. Because of poor heat dissipation and the absence of any lubricant exchange, the whole of the working film spends all its time at or near the maximum film temperature.

The final category of heat input is that of flash temperatures. At the same speeds and loads the flash temperatures with a grease are likely to be similar to, or greater than, those occurring with oil. However, because the quantity of lubricant is limited to the thin film on the moving surfaces, the whole of the degradation caused by the flash temperatures will be retained in that small quantity of grease.

It was estimated in Chapter Seven (Table 4) that the oil in a self-contained bearing spends about 10^{-5}–10^{-4} per cent of its total operating time exposed to flash temperatures. The grease film in a similar situation will probably spend 0.01–0.1 per cent of its total time experiencing flash temperatures, because it is neither diluted nor replaced with fresh grease.

The overall thermal stress on a grease is therefore much greater than on an oil in a similar bearing situation, because the cooling with grease lubrication is poorer, and because of the aggravated effect of flash temperatures. It follows that even if the thermal and oxidative stability of the grease are as good as those of the oil, deterioration of the grease will take place more quickly.

In practice, however, grease-lubricated components cannot be operated as fast as oil-lubricated ones. The poorer heat removal with greases leads to higher bearing temperatures, and in extreme cases these higher temperatures will be high enough to cause rapid breakdown of the grease. Grease lubrication is therefore limited to tangential speeds of about 2 m/s for plain bearings or gears, or 20 m/s for rolling bearings, compared with 300 m/s for oil.

These effects are in addition to the temperature limits imposed on greases by the stability or instability of the gel structure, which will be discussed later.

Several standard procedures exist for the prediction of grease life, such as *Design Data Item 78032* published by the Engineering Sciences Data Unit. They generally recommend the application of an exponential life reduction factor such as 0.658 for every 10°C (ESDU) or 0.5 for every 15°C (Gesellschaft für Tribologie) above 70°C. This is based on a standard such as lithium soap-thickened mineral oil, and a different reference temperature would probably apply to other types of grease.

The one advantage of a grease for high-temperature applications is that its consistency changes relatively little as the temperature increases, until it approaches the Drop Point. Figure 43 shows how the penetration of a sodium soap grease is less than doubled as the temperature rises from 20°C to 180°C. Over the same temperature range the viscosity of an oil might decrease by a factor of 20 or 30.

This is an advantage in ensuring the retention of the grease adjacent to the moving parts, and in maintaining the lubricant film thickness in low-

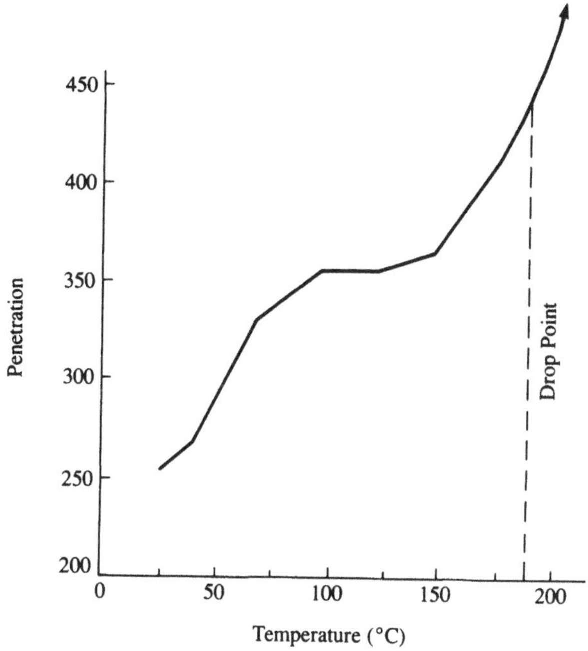

Fig. 43 Variation of penetration with temperature for a sodium soap grease

speed components. In high-speed components the effect of shear reduces the effective viscosity until it approaches that of the base oil, which is affected by temperature in the same way as with any oil.

In addition to the base oil and thickener, greases may contain a wide variety of other components, and Table 9 lists a wide range of base oils, thickeners, and additives for greases. Most of the additives are similar to those used in oils, but two which are specific to greases are structure modifiers and water repellants.

Lubricating pastes differ from greases in that they consist of dispersions of a solid in a liquid, without any stable gel structure. As a result the solid component can separate out under gravity or under gentle centrifuging. Their behaviour in moving components is different from that of greases, and they can only be used satisfactorily in low-speed systems. There are also pastes which consist of solids dispersed in grease or petrolatum. These are more stable but still have poor flow properties, and are most often used as anti-seizes in static assemblies such as threaded joints.

Table 9 Some components used in grease formulation

Base oils	*Thickeners*	*Additives*
Mineral oils	Sodium soap	Anti-oxidants
Synthetic hydrocarbons	Calcium soap	Anti-wear additives
Di-esters	Lithium soap	EP additives
Polyol esters	Aluminium soap	Corrosion inhibitors
Silicones	Barium soap	Friction modifiers
Phosphate esters	Aluminium complex	Metal deactivators
Fluorinated silicone	Calcium complex	VI improvers
Chlorinated silicone	Sodium complex	Pour-point depressants
Perfluoroalkyl ethers	Lithium complex	Tackiness additives
	Bentonite clay	Water repellants
	Silica	Dyes
	Carbon/graphite	Structure modifiers
	Polyurea	Solid lubricants
	PTFE	
	Polyethylene	
	Indanthrene dye	
	Phthalocyanine dye	

8.2 DROP POINT

When the temperature of certain greases is raised, a point is reached at which the gel structure is destabilized and the grease becomes liquid. This temperature is called the Drop Point or Dropping Point of the grease.

Drop Point is defined as the temperature at which a drop of grease or other petroleum product first detaches itself from the main bulk of material when a sample is steadily heated under prescribed conditions. It is an indication of the temperature at which a grease changes from the solid or the semi-solid to the liquid state. It is measured by one of several standard tests, such as ISO 2176, IP132, and ASTM D566.

The Drop Point represents the highest temperature at which a grease can remain a grease. If a grease is heated above its Drop Point and then cooled, it will in most cases return to a grease-like consistency, but this is likely to be softer than its original consistency. Since a grease's performance in many respects depends critically on its consistency, the change represents an important deterioration. It should therefore be assumed that once a grease has been heated above its Drop Point it is no longer suitable for use.

There are two possible exceptions to this. The first relates to flash temperature. If the grease in a lubricated contact is exposed to flash

Table 10 Drop Points of certain types of grease

Thickener type	*Drop Point* (°C)
Calcium soap	60–100
Aluminium soap	100–120
Barium soap	120–150
Polypropylene (atactic)	140–150
Sodium soap	165–190
Lithium soap	170–200
Aluminium complex	200–260
Calcium complex	225–250
PTFE	230–300
Sodium complex	240–260
Lithium complex	250–320

temperatures higher than its Drop Point, the volume of grease involved is likely to be so small that the consistency of the grease film as a whole is not likely to be affected.

The second case relates to the temperature in the lubricating film as a whole. The grease in the thin film, especially at high shear rates, no longer retains its grease structure. It follows that the Drop Point, which is both a function of, and an effect on, the gel structure, is probably of little importance in the thin lubricating film in high-speed bearings.

In low-speed components, however, the grease maintains a gel structure, even if it is modified by mechanical shear, and the maintaining of adequate film thickness depends to some extent on the consistency of the grease. If the lubricating film is heated to the Drop Point in such a bearing, there is likely to be a loss of performance.

This case may be only academic, because in a low-speed component the film temperature is unlikely to rise much above the bulk temperature.

The Drop Point is almost entirely related to the thickener used in a grease, and Table 10 lists the Drop Points of greases with some of the commoner types of thickener.

Some non-soap greases do not have a Drop Point. In practice this means that some other type of deterioration determines the maximum temperature of use. An intermediate example is that of the calcium complex greases. They have Drop Points of 250–300°C, but the soap component begins to decompose at about 140°C.

However, this creates an important distinction, because the temperature limit due to Drop Point is absolute and not time-dependent, whereas decomposition effects are time-dependent. In practice, the maximum operating bulk temperature is determined by the Drop Point for short periods of operation, but over long periods is limited by decomposition.

8.3 THERMAL STABILITY

The thermal stability of greases depends on the individual components present. The stability of the base oils has already been discussed in Chapters Four and Seven, and that of some of the boundary additives in Chapter Six.

Soap thickeners are organic compounds similar in structure and properties to the long-chain acids mentioned in Chapter Six. They begin to decompose significantly at temperatures between about 120 and 220°C. The decomposition is time-dependent, and would presumably follow logarithmic patterns similar to those presented for the various oils in Chapter Seven, but the available information is not sufficient to construct similar life–temperature graphs.

Some of the non-soap thickeners are thermally stable to very high temperatures. An extreme example is silica, which is stable at temperatures over 1500°C. Some of the modified clays are very stable in themselves, but when used as grease thickeners they often rely on the presence of organic chemicals adsorbed on their particles, so that, in practice, their stability is limited by those chemicals. Even so, some of them are stable enough even at 300°C to give useful grease lives.

Although the inherent thermal stability of a substance cannot be improved by additives, it can be decreased by other substances. Some greases are a complex mixture of different materials, and the greater the variety of substances present the greater is the risk of some interaction reducing the overall thermal stability.

8.4 OXIDATION

As with thermal stability, the oxidative stability of greases is mainly determined by the individual oxidative stabilities of the components.

Where the base oil is a mineral oil or di-ester, the overall oxidative stability of the grease is determined mainly by the stability of the oil, and follows Figs 20 and 26. Where the more stable high-temperature base oils are used, it is necessary to use the more stable thickeners, such as silica,

PTFE, and some of the high-temperature indanthrene or phthalocyanine dyes. As with the oils, it is difficult to find satisfactory anti-oxidants, anti-wear additives, or EP additives for use above about 260°C.

One useful EP additive for high-temperature greases is molybdenum disulphide. When used in oil, molybdenum disulphide presents a problem because the very fine particles which will form stable dispersions in an oil are not ideal for lubrication. In greases coarser particles of molybdenum disulphide will remain fully dispersed, and can provide exceptionally-high load-carrying capacity. Molybdenum disulphide has very useful oxidation resistance up to 350°C.

When organic substances are oxidized, one of the first and most common results is acidity, and this has a de-stabilizing effect on the gel structure of most greases. This effect is due to the fact that many gel structures are stabilized by an electrical double layer, which tends to be neutralized in the presence of the more conductive acid products.

As a result, even mild oxidation tends to produce a softening of greases. Further oxidation then leads to darkening, and eventually to hardening.

The lower rate of diffusion of oxygen into a grease gives some protection against oxidation for grease in bulk in the unswept parts of a component. The thin lubricating film has no such protection. It is in close contact with oxygen if air is present, the oxygen is well dispersed in the film by mechanical shearing, the film is heated by frictional energy, and anti-oxidants become rapidly depleted because re-supply is limited.

The result is that the working film of the grease is oxidized rapidly at high temperatures. As a rough guide the endurance of a grease subject to oxidation is probably only about 10 per cent of the life of the equivalent oil at the same temperature.

Replenishment of grease at high temperatures is therefore particularly important, but is much more difficult to achieve than for oils.

8.5 TEMPERATURE LIMITS FOR GREASES

As a generalization, the temperature limits for a grease must be defined in two parts. If there is a Drop Point, this will be an absolute limit which is not time-dependent. Below the Drop Point, the limit will be time-dependent, and the life will vary with temperature, as shown previously for oils, except that the life of a grease at any temperature will be shorter than for the equivalent oil.

The limits for specific grease types in the following sections are subject to all the uncertainties which were discussed for oils in Chapter Seven, and should again be considered only as rough guides.

8.6 MINERAL OIL GREASES

The vast majority of lubricating greases are mineral oil based, and most have either soap or complex thickeners. They can be obtained in all NLGI grades from the softest No. 000 (penetration 445–475) to the hardest No. 6 (penetration 85–115), and in many specialized versions such as graphite (conducting), molybdenum disulphide (high load-carrying or low-friction), high-tack, and so on.

Their temperature limits depend mainly on the type of thickener used, and typical limits are shown in Fig. 44.

8.7 SYNTHETIC HYDROCARBON GREASES

Greases based on synthetic hydrocarbon oils are used in relatively small quantities for special applications, such as in aircraft, where use over wide ranges of temperature is necessary. It may, therefore, be easier to obtain them from aviation suppliers.

Because they are commonly used at high temperatures, only the higher-temperature thickeners, such as lithium and calcium complexes, are normally used. Figure 45 shows typical temperature limits for complex-thickened synthetic hydrocarbons.

8.8 DI-ESTER GREASES

Di-ester greases were probably the first greases based on synthetic oils to be used in significant quantities. Their principal use was initially in aircraft, but they have more recently been quite widely used for higher-temperature industrial applications, while they have been replaced by synthetic hydrocarbons for some aviation applications. The NLGI grades available are usually limited to No. 2 (penetration 265–295) or No. 3 (penetration 220–250).

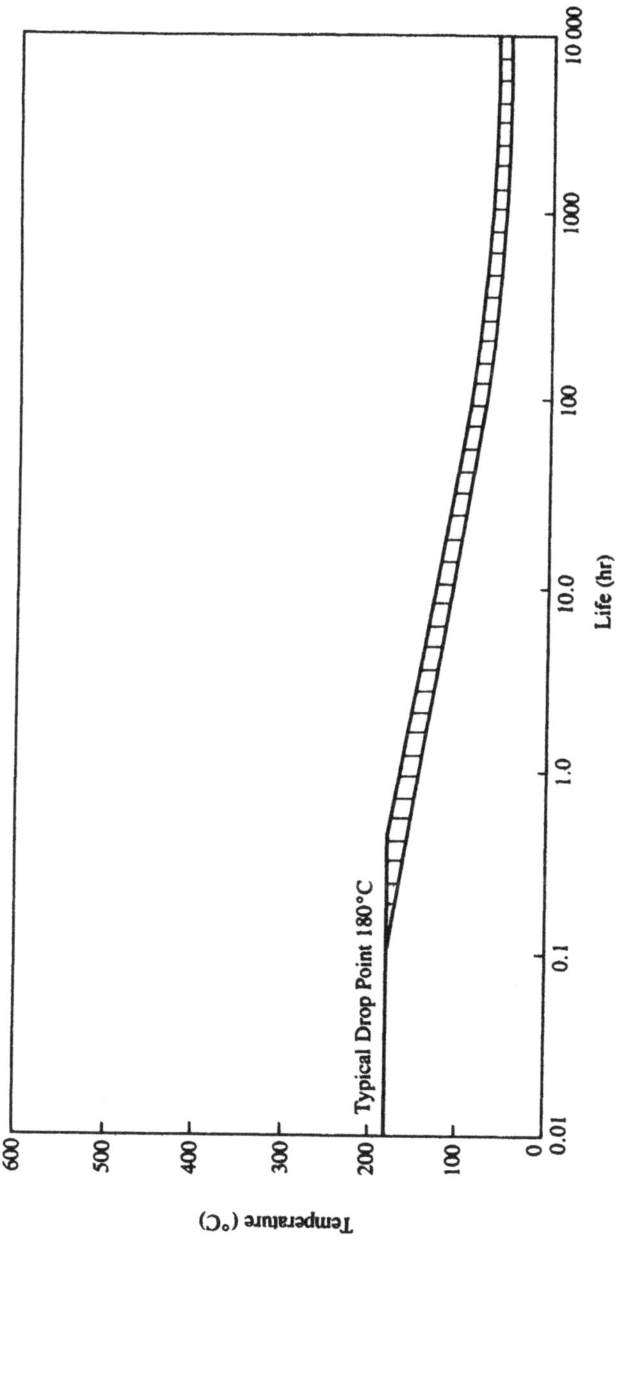

Fig. 44 Temperature limits for mineral oil greases

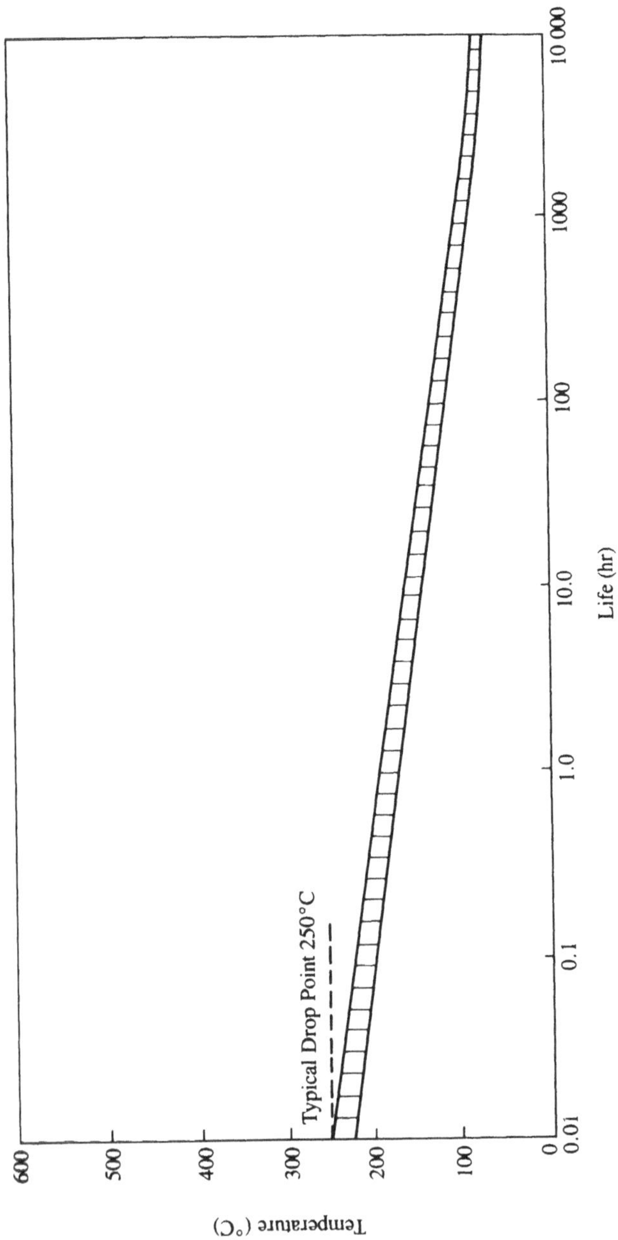

Fig. 45 Temperature limits for synthetic hydrocarbon greases

The earliest di-ester greases were usually thickened with lithium soaps or bentonite, but more recently they often use lithium complex thickeners. Typical temperature limits are shown in Fig. 46.

8.9 POLYOL ESTER GREASES

Polyol ester greases are used in limited quantities for applications at temperatures higher than those for which di-esters or synthetic hydrocarbons are used. The grades available are probably limited to No. 2 and No. 3. Typical temperature limits are shown in Fig. 47.

8.10 POLYGLYCOLS

Polyglycol greases are of two general types, depending on whether they are designed for clean burning or for hydrocarbon-insolubility.

Greases for clean-burning can use either ether or diol polyglycols. They are intended for use in applications, such as conveyor chains in ovens or furnaces, where the temperatures are significantly higher than those at which polyglycols can survive. The principle is that the polyglycol decomposes cleanly, leaving no residue. Solid lubricants, usually graphite or molybdenum disulphide, are added to provide further lubrication once the polyglycol has decomposed.

Lithium soap may be used as the thickener. This leaves a small deposit on decomposition, although this tends to be powdery so that it causes little difficulty. A better thickener is a good quality graphite, which provides useful lubrication after the decomposition of the polyglycol, and in turn decomposes to leave little or no solid residue.

Greases insoluble in hydrocarbons are useful for lubricating seals, packed glands, and valves in pumps for hydrocarbon fuels and refinery or petrochemical plant product systems. They are based on diol polyglycols, but often with some proportion of ether-terminated polyglycol to provide better low-temperature fluidity, thickened usually with lithium soap, and are not intended for high-temperature use.

8.11 PHOSPHATE ESTER GREASES

Greases can be formulated from several different types of phosphate ester. They usually have lithium soap thickeners because the high

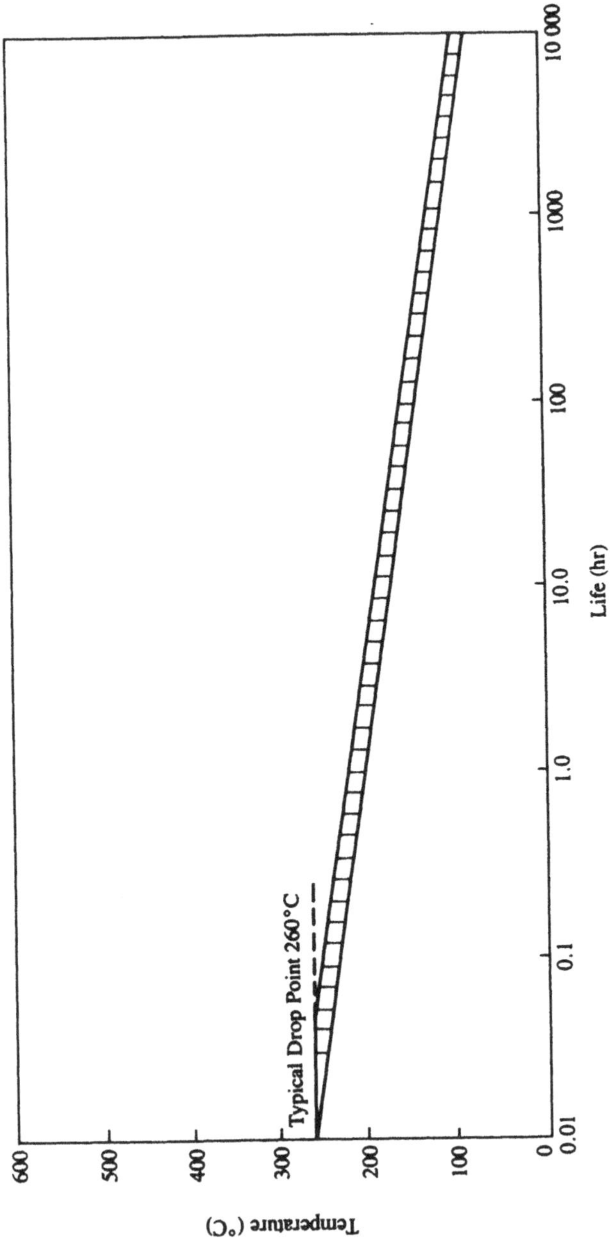

Fig. 46 Temperature limits for di-ester greases

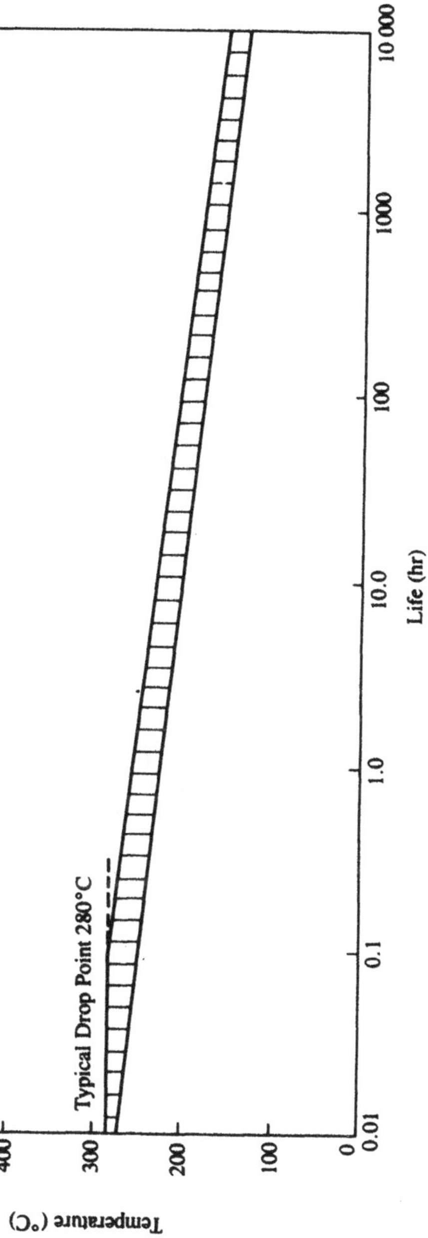

Fig. 47 Temperature limits for polyol ester greases

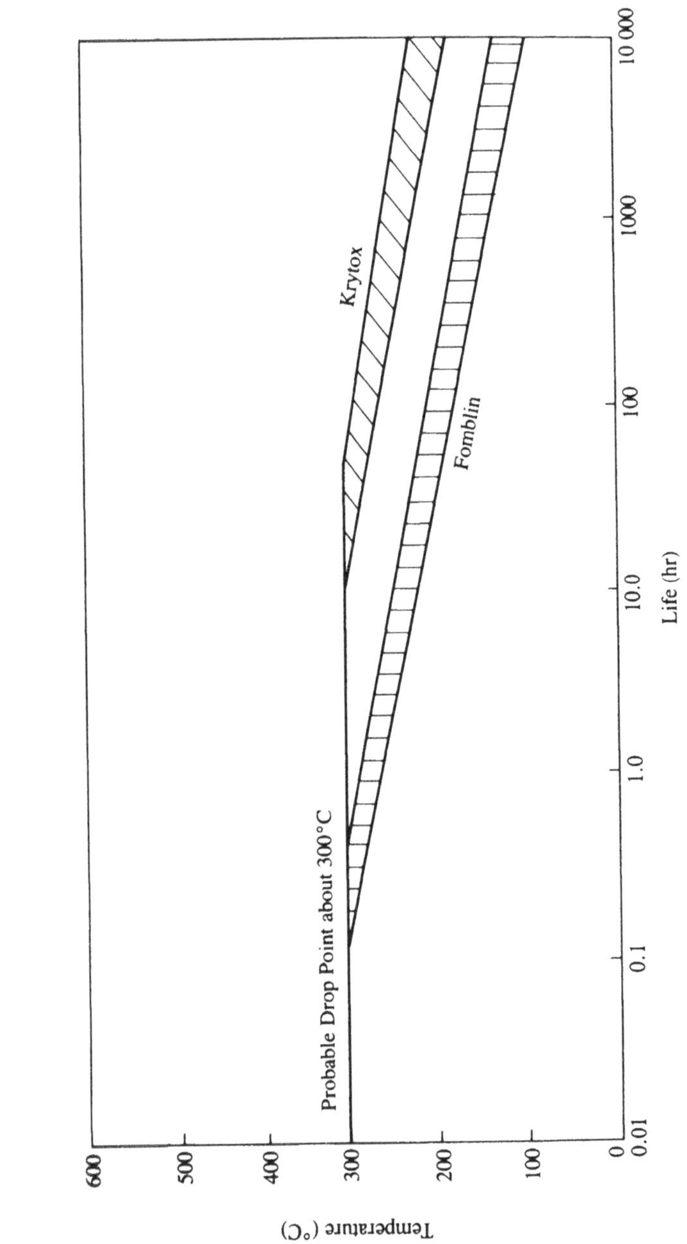

Fig. 48 Temperature limits for perfluoropolyalkyl ether greases

temperature performance of phosphate esters is mediocre. The greases are used almost entirely in locations with very high fire hazards because of their outstanding fire resistance. They are good lubricants and have been claimed to be stable up to over 200°C, but their life at such temperatures would be very limited.

8.12 SILICONE GREASES

Silicone greases were among the earliest greases based on synthetic oils. They can be thickened with lithium soap, but the Drop Point of the resulting grease is low compared with the temperature capabilities of the silicone fluids. Better high-temperature performance is obtained with non-soap thickeners such as indanthrene blue dye, copper phthalocyanine, carbon black, and silica. The last two do not have Drop Points, and all of these high-temperature greases can be used to temperatures between 230 and 310°C.

Like the silicone oils, silicone greases are rather poor lubricants for steel against steel, especially in plain bearings. In ball or roller bearings they are generally satisfactory, provided that the bearing loads are kept down to about 30 per cent of their normal rating, and operating speeds are relatively low. Molybdenum disulphide is often added to silicone greases to improve lubrication.

8.13 PERFLUOROPOLYALKYL ETHERS

Excellent greases can be made from perfluoropolyalkyl ethers by thickening with fine PTFE powder. They are quite good lubricants and stable to about 300°C. There has been little published information about their operating life at different temperatures or their performance in bearings because of their very high cost. Figure 48 shows an estimate of their temperature limits.

8.14 OTHER GREASES

Greases have been made from silicate esters and used to a limited extent in weapons. Greases have also been made experimentally from polyphenyl ethers, chlorofluorocarbons, ferrocene derivatives, and silanes, but it is not clear that any of them has ever been made available commercially.

CHAPTER NINE

Solid Lubricants and Composites

9.1 GENERAL

From the previous two chapters it is clear that there are severe limitations on the use of liquid-based lubricants at high temperatures. It is therefore interesting to consider the alternative of using solids.

Compared with oils or greases, solid lubricants have a number of disadvantages, and they are not used in large amounts at normal temperatures. The total use of solid lubricants is less than 0.5 per cent of that of liquids and greases. However, for high temperatures some of them have very important advantages, including high chemical stability, high melting point, low volatility, the absence of viscosity effects, and a general absence of any chemical reaction with other materials. The main advantages and disadvantages of solid lubricants are listed in Table 11, but these will vary between different solids.

Any solid material which can reduce friction when interposed between two surfaces can act as a solid lubricant, but to be really effective it should preferably have several other properties, such as the following.

(1) It should adhere, or be made to adhere, to one or both of the bearing surfaces so that it is not rapidly lost from the mating surfaces.
(2) It should provide low but consistent friction between two surfaces.
(3) It should be chemically stable over the required temperature range.
(4) It should not attack or damage the bearing materials.
(5) It should have sufficient wear resistance to provide a useful life.

The major disadvantages which need to be highlighted whenever consideration is given to the use of solid lubricants are the unavoidable continuing wear and limited wear life, the virtual impossibility of relubrication in operation, the higher friction, and the poor heat transfer.

In relation to the four categories of heat input discussed in Chapter Two, solid lubricants are influenced only by ambient temperature, component temperature, and flash temperatures. System temperatures, as defined previously, are usually not relevant because there is rarely, if ever, any feed system present.

The effect of ambient temperature is similar to its effect on oils or

Table 11 Advantages and disadvantages of solid lubricants compared to oils and greases

Advantages

(1) Practically no tendency to flow, creep or migrate, so that they can be relied on to remain in place for long periods.
(2) Little tendency to contaminate adjacent systems or products.
(3) Very low volatility makes some of them usable in high vacuum.
(4) Usable at very high or very low temperature.
(5) Chemical inertness to other chemicals.
(6) Generally more stable to radioactivity.

Disadvantages

(1) Difficult or impossible to feed or replenish.
(2) Finite wear rate limits their life.
(3) Poor thermal conductivity limits sliding speed.
(4) Thermal expansion often very different from metals, so that clearances can be lost with temperature change.
(5) Generally higher friction.
(6) More complicated to apply.
(7) No coolant properties.
(8) No corrosion protection.

greases, but because of the high temperature stability of most solid lubricants, much higher ambient temperatures are acceptable.

Component temperatures tend to be higher in relation to ambient temperature for solid lubricants for the same loads and speeds because of the higher friction. Because of their poor heat transfer, the temperature gradients also tend to be greater, so that lubricant film temperatures are high. Thus the superior thermal and oxidative stability of solid lubricants is partly offset by the higher temperatures generated within the lubricant film. The greatest temperature advantage is at low speeds.

High loads can be carried without the same heating tendency. The power dissipation in the bearing is directly proportional to the load, the speed, and the coefficient of friction.

$$P = \mu WV \qquad (56)$$

However, the coefficient of friction for many solid lubricants is not constant, but increases as the speed increases and decreases as the load increases. It follows that the power dissipation is less than proportional to the load but more than proportional to the speed.

As a result, the sliding speed which is possible with solid lubricants is severely restricted. The maximum speed varies with the type of lubricant, the thickness of the lubricant layer, the thermal properties of the support structure, any cooling effects, the contact area, and the load, but as a rough general rule the maximum permissible sliding speed is about 0.5 ms^{-1}. Metallic solids can sustain higher speeds because of their higher thermal conductivity, and some improvement can be obtained with other lubricants by incorporating metals in them.

It was shown in Chapter Eight that the effect of flash temperatures is proportionately greater for a grease than for an oil, because of the lower scope for thermally degraded material to be dispersed in an adjacent volume of lubricant. In the case of a solid lubricant there is virtually no scope for dispersing degraded lubricant, except by wear, which should be a relatively slow process. It follows that all thermal or oxidative degradation of a solid lubricant leads directly to a deterioration of the actual contact surface of the lubricant and, therefore, to a deterioration in lubrication.

The best solid lubricants for use in air all have poor thermal conductivities, while their coefficients of friction are higher than for oils or greases. As a result, flash temperatures tend to be high, and calculations suggest values as high as 1000°C. Material which is thermally or oxidatively degraded by these high flash temperatures is retained at the surface, and is progressively further degraded by repeated exposure to flash temperatures.

The practical effect of this continual exposure to high film temperatures and flash temperatures depends on the class of solid lubricant involved, and will be described in more detail in the following sections.

The solid nature of solid lubricants leads to many different features in their use, compared with oils or greases. The one which is perhaps most obvious is the difficulty, or even impossibility, of re-supplying a solid lubricant to a component to replace the initial supply when it reaches the end of its useful life.

A second important feature which is very much to the advantage of solid lubricants is their virtually infinite storage lives at quite high temperatures. Unless the solid is unusually porous, oxidation normally takes place only on the surface, and produces a very thin degraded film which has little adverse effect on subsequent performance. This is of considerable value in equipment such as safety valves, other safety equipment, security alarms, or guided missiles, which may be required to survive long inoperative periods and subsequently operate with very high reliability under hostile conditions.

There is of course a limit to the temperature which a solid lubricant will survive in storage. Excessive temperatures will lead to gross oxidation or other forms of deterioration. Nevertheless, most solid lubricants can survive for very long periods temperatures much higher than normal ambient temperatures.

Solid lubricants can be used in many different forms, including the following.

(1) As free powder.
(2) Applied in suspension in a volatile liquid.
(3) Burnished onto a surface by rubbing.
(4) Applied as a coating in an organic resin, polymer, metal or inorganic binder.
(5) Plasma-sprayed coating.
(6) Sputtered coating.
(7) Ion-beam assisted coating.
(8) Ion implantation.
(9) In-situ formation.
(10) Incorporation in composites.

They are also incorporated in oils or greases, but in that form they comprise additives rather than solid lubricants.

It is not the object of this chapter to provide detailed guidance on the procedures for using solid lubricants, but only to describe factors relating to high-temperature use. It should, however, be recognized that solid lubricants differ very considerably from oils or greases, in terms of application techniques, performance, and design. Specialist guidance should always be obtained before using solid lubricants in any application for the first time.

The best application technique depends on the specific lubricant and the nature of the component, and should be discussed with the suppliers or other specialists.

Solid lubricants can be broadly divided into four categories, which are listed in Table 12. The whole range of solids which can be used as lubricants includes scores of compounds, and many of these have been studied in composites containing two, three, or more components in different proportions. The result is that hundreds of different solid lubricants have been reported, and it is impossible to describe them all here. The range described will therefore be limited to a few of the most important individual compounds, and composites will be described later in the chapter.

Table 12 Some successful solid lubricants

Layer-lattice compounds	
Molybdenum disulphide, MoS_2	Graphite, C
Tungsten diselenide, WSe_2	Tungsten disulphide, WS_2
Niobium diselenide, $NbSe_2$	Tantalum disulphide, TaS_2
Calcium fluoride, CaF_2	Graphite fluoride, CF_x
Barium fluoride/calcium fluoride	
Polymers	
Polytetrafluoroethylene, PTFE	Nylon
Polytrifluorochloroethylene, PTFCE	Acetal
Fluorinated ethylene-propylene, FEP	Polyimide
Polyvinyl fluoride, PVF_2	Polyphenylene sulphide
Ultra-high molecular weight polyethylene	Phenolics
Polyetheretherketone (PEEK)	
Metals	
Lead	Silver
Tin	Gold
Indium	Barium
Other inorganics	
Molybdic oxide, MoO_3	Boron trioxide, B_2O_3
Lead monoxide, PbO	Boron nitride, BN
Lead sulphide, PbS	Tungsten trioxide, WO_3
Barium oxide, BaO	Cadmium oxide, CdO

9.2 LAYER-LATTICE COMPOUNDS

Some of the most effective solid lubricants, including molybdenum disulphide and graphite, have a crystal structure known as a layer-lattice structure. This consists of thin flat parallel layers, usually with an internal hexagonal structure. Interatomic forces are strong within the flat layers, but very weak between the layers, so that easy sliding takes place between the thin layers, or platelets.

Ideally the lubricants adhere to the bearing surfaces with the flat platelets oriented parallel to the surfaces. The strong forces in the plane of the platelets then ensure good coating integrity, while the weak van der Waals forces between the layers give easy sliding and low friction. Friction is further reduced if some of the platelets transfer to the sliding counterface, so that all sliding takes place between lubricant platelets and not between bearing surface and lubricant.

Films as initially deposited by many of the available techniques are soft and un-oriented, and in that condition show high friction and high wear rates. By careful burnishing or running-in, the coating is densified, giving the required orientation and considerably reducing friction and wear.

This consolidation process is not appropriate for sputtered, ion-implanted, or in-situ coatings, but as a result these have less perfect orientation.

9.2.1 Molybdenum disulphide and similar compounds

Several disulphides, diselenides, and ditellurides of molybdenum, tungsten, niobium, and tantalum have very similar crystal structures and lubricant performance. They are known collectively as lubricating dichalcogenides.

The only one of the group which occurs naturally is molybdenum disulphide, which exists as the mineral molybdenite, the main ore of molybdenum. The others are all manufactured synthetically, and as a result they are more expensive and less widely used. The technology of molybdenum disulphide is therefore much more fully developed.

Molybdenum disulphide can be applied by all the techniques listed previously. At normal temperatures it is most often used in the form of bonded films. These are prepared by dispersing molybdenum disulphide powder in a liquid or semi-liquid material such as an organic resin. The dispersion is applied to a surface in the form of a thin film by dipping, brushing, or, preferably, spraying, and then dried or cured to give a solid film. Such films in their original dried or cured form are often too soft, and need to be densified by running-in or by burnishing. For high-temperature applications the best binders are inorganic, and several such coating materials are available commercially.

In its fully-densified form it has a coefficient of friction as low as 0.02 at very high contact pressures, increasing to between 0.04 and 0.08 at about 0.7 GN/m^2, and then remaining fairly constant at lower pressures. Higher coefficients of friction have been reported, up to about 0.45, but it seems probable that the highest of these refer to composites or contaminated or poorly-oriented films.

The low coefficient of friction of molybdenum disulphide is an inherent property, and is adversely affected by most contaminants, including moisture.

In an inert atmosphere or high vacuum the coefficient of friction remains low up to well over 900°C. At higher temperatures in vacuum there is some loss of weight due to evaporation, and decomposition to molybdenum and sulphur takes place at an increasing rate as the temperature rises from 1000°C to the melting point at about 1700°C. It seems probable that molybdenum disulphide can be used satisfactorily in the absence of oxygen for extended periods at 1000°C and for short periods at 1500°C.

In the presence of oxygen the life of most molybdenum disulphide coatings is limited by oxidation rather than wear. It has been variously reported that an increase in friction due to oxidation begins at some temperature between 300°C and 480°C. In fact such a statement is meaningless. Because of the high flash temperatures which occur with solid lubricants, some oxidation will inevitably take place at moderate loads and speeds even at normal ambient temperatures. Apart from the minor effect of wear, the oxidized material will remain in the surface, with a progressive increase in friction until the amount of oxide present is sufficient to cause the coating to break up. Certainly, useful lives can be obtained at temperatures up to 450°C in air.

The wear rate of a fully-consolidated film is very low. Figure 49 is a nomograph which gives some guidance on wear life of molybdenum disulphide at normal temperatures, but there is insufficient information available to provide similar guidance at higher temperatures.

Tungsten diselenide and tungsten disulphide are less affected by oxidation than molybdenum disulphide up to some temperature between about 480°C and 600°C. This is believed to be due to the fact that the initial film of tungsten trioxide formed on their surfaces gives better protection against further oxidation than is provided by molybdic oxide.

The synthetic lubricating chalcogenides appear to have little other advantage over molybdenum disulphide, and none of them has been very extensively used.

9.2.2 Graphite

The structure of graphite is very similar to that of molybdenum disulphide, but when it is completely pure the interatomic forces between the crystallite layers are much higher, so that the friction is also high. Low friction is only obtained when certain contaminants such as moisture are present.

Historically, graphite has been more widely used than molybdenum disulphide, but the latter has been more extensively studied in recent years. Graphite is still used as a free powder, in dispersions in volatile

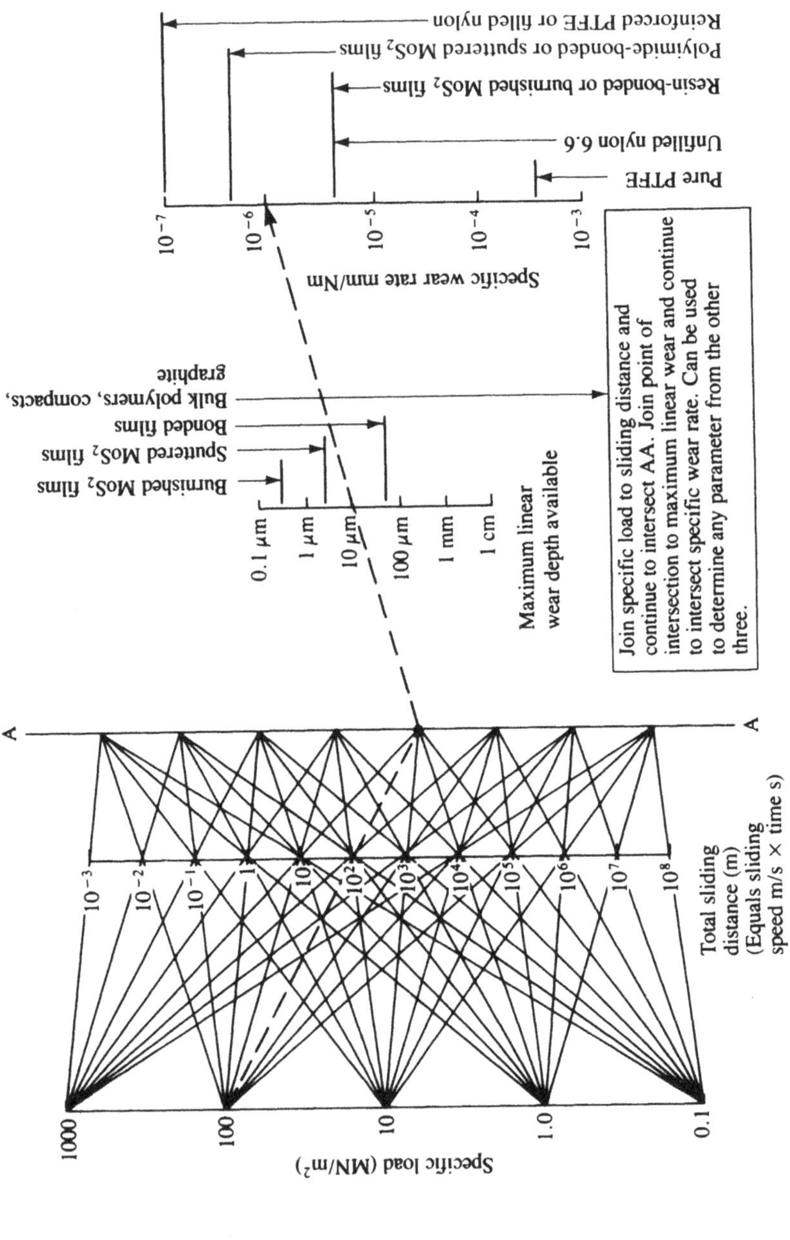

Fig. 49 Selection of solid lubricant type

solvents, and in oils and greases, but the other more sophisticated application techniques have been little used for it, if at all.

The coefficient of friction varies between 0.05 and 0.15, increasing as the pressure decreases. As the temperature rises in normal atmospheres the usual moisture film is driven off and becomes inadequate at about 200°C. The friction then rises to about 0.3 or 0.4. Some improvement has been obtained by adsorbing other vapours onto the graphite, but these again become ineffective as the temperataure rises past their desorption temperatures.

The mechanism by which these adsorbents decrease friction is believed to be 'intercalation' between the crystal layers, which reduces the interatomic forces and therefore the shear stress. Deliberate intercalation has been studied with compounds like nickel chloride, which helps to maintain low friction in the range 200–320°C, but the resulting friction was no lower than 0.17.

There is therefore a gap between about 200°C and 350°C in which graphite is a mediocre lubricant. Above 350°C it again becomes an effective lubricant, and it has been suggested that this may be due to some interaction with metal oxides. At about 500–550°C oxidation becomes rapid in air, and graphite becomes unusable. In some applications it is an important advantage that graphite oxidizes to form gaseous carbon dioxide, so that no undesirable residue is formed.

Graphite cannot be used in vacuum because of the absence of the adsorbed moisture or other vapours needed to maintain low friction. There seems to be no reason why it could not be used in inert atmospheres to high temperature if some means could be found for maintaining the necessary adsorbed films, but graphite offers insufficient advantage over molybdenum disulphide and some of the other lamellar solid lubricants for the attempt to be justified.

9.2.3 Calcium fluoride and barium fluoride

Calcium fluoride is an effective lubricant between about 400°C and 900°C. In the solid state it will not adhere sufficiently to bearing surfaces, but it can be made to adhere by melting ('fusion-bonding'), binding in a ceramic by firing, or incorporation in a high-temperature polymer such as a polyimide. The coefficient of friction in a fusion-bonded film is in the range 0.1–0.15.

Barium fluoride can be used in similar ways, but the two materials seem to be best in the form of a eutectic containing 38 per cent calcium fluoride and 62 per cent barium fluoride. The eutectic has, of course, a lower melting point than either of the components, and this is an advantage for fusion-bonding. The coefficient of friction is between 0.1 and 0.2 at temperatures between 250°C and 800°C, but rises to 0.4 at normal temperatures. The problem of operating at low temperature before a machine reaches a high equilibrium operating temperature has led to a search for other components to bridge the low-temperature gap, and one resulting composite is described later.

9.2.4 Graphite fluoride

Graphite can be treated with fluorine to produce a lamellar crystalline solid in which the atomic ratio of graphite to fluorine can vary from 4:1 to 9:10. With the higher proportions of fluorine the product is white, and the coefficient of friction lies between about 0.03 and 0.15 over the temperature range from normal temperature to 470°C. Above 470°C graphite fluoride decomposes, and presumably the proportion of fluorine falls, leading to higher friction.

9.3 POLYMERS

Many different polymers are used as solid lubricants, but only PTFE is really effective or widely used. Table 13 lists ten of the more widely used

Table 13 Properties of some polymeric solid lubricants

Polymer type	*Limiting temperatures (°C)* Long periods	Brief use	*Friction coefficient*
Polytetrafluoroethylene (PTFE)	250	320	0.03–0.1
Polytrifluorochloroethylene (PTFCE)	200	210	0.07–0.3
Fluorinated ethylenepropylene (FEP)	205	220	0.06–0.15
Polyvinyl fluoride (PVF_2)	150	175	0.1 –0.25
Nylon	120	150	0.05–0.8
Acetal	85	110	0.15–0.35
Polyetheretherketone (PEEK)	150	250	0.1 –0.3
Polyimide	260	480	0.1 –0.25
Polyphenylene sulphide (PPS)	230	260	0.1 –0.3
Ultra-high molecular weight polyethylene	80	95	0.3 –0.5
Phenolics	110	150	0.1 –0.2

basic types with an indication of their temperature limits and coefficient of friction.

PTFE is the only one of the ten which is used in the form of thin coatings. In the pure state it is rather soft, and some form of reinforcement is necessary to increase its wear life. Unfortunately all the reinforcement techniques result in an increase in friction, usually to at least 0.1–0.15.

All the other polymers listed are used in dry sliding applications only at relatively low loads and speeds. Most are more commonly used in marginally-lubricated situations, where low-viscosity liquids such as water, fuels, or solvents are present, or where there are traces of oil or grease.

Polymers in general will not normally give very high unlubricated friction against metals, but their friction is high enough to cause significant surface heating and softening. They are commonly incorporated in composites, and these are described later.

9.4 METALS AS SOLID LUBRICANTS

Certain metals have low coefficients of friction in dry sliding against metallic bearing surfaces, and their high thermal conductivity is very useful in dissipating frictional heat. The six most commonly used are listed in Table 14 with their melting points and coefficients of friction.

Flash temperatures with metals are lower than for the layer-lattice compounds and the other inorganic compounds described later, but they are high enough to cause oxidation of all but gold. This imposes a major limitation on the use of metals as solid lubricants in air.

Some of the lubricating metals are very useful as solid lubricants in high

Table 14 Some metals used as solid lubricants

Metal	*Melting point (°C)*	*Coefficient of friction*
Barium	725	
Gold	1064	0.07 –0.11
Indium	157	0.04 –0.2
Lead	328	0.015–0.7
Silver	962	0.09 –0.3
Tin	232	0.02 –0.6

vacuum. Lead has been found particularly effective for the lubrication of rolling bearings in space mechanisms. Gold, silver, and barium have also been used in space, as well as in such equipment as electron microscopes and X-ray equipment.

Soft metals must be used in thin films on hard substrates to provide strength under load, but the friction depends critically on the film thickness. The optimum film thickness is usually between 0.1 and 1.0 μm, and is obtained by plating, vacuum deposition, sputtering, or plasma deposition. It has been found desirable for the surfaces of the substrate and the film to be very smooth and uniform.

At low speeds and loads the lubricating metals can be used at temperatures close to their melting points, but at higher speed and load frictional heating may cause local melting. Asperity melting due to flash temperatures has no adverse effect, and can have three beneficial effects. These are reduction in friction, smoothing of the surface, and lowering of the flash temperature due to the removal of the latent heat of fusion.

9.5 OTHER INORGANIC SOLID LUBRICANTS

A wide variety of inorganic compounds has been investigated over many years for potential use as solid lubricants for high temperatures. Eight of the best-studied are listed in Table 15 with some of their important properties.

None of these compounds seems to have been used in any important service applications yet. The exact reason for this is not clear, because

Table 15 Some inorganic solid lubricants

Lubricant	*Temperature range* (°C)	*Coefficient of friction*
Molybdic oxide, MoO_3	300–1000	0.16–0.25
Lead monoxide, PbO	200–700	0.06–0.09
Lead sulphide, PbS	300–550	0.2 –0.3
Barium oxide, BaO		0.06–0.12
Boron trioxide, B_2O_3	550+	0.1
Boron nitride, BN	300–950	0.2 –0.3
Tungsten trioxide, WO_3		0.2
Cadmium oxide, CdO		0.05–0.11

research workers have tended always to report their results in optimistic terms.

Lead oxide has quite low and consistent friction over a wide temperature range, being below 0.1 over the whole range from 200°C to 650°C, but not only has it not been much used as such, but its formation is believed to be the major problem preventing the use of lead in air.

The others all have rather higher friction and the disadvantage that they do not readily adhere to bearing surfaces.

9.6 SELF-LUBRICATING COMPOSITES

All solid lubricants have disadvantages, even when used in the conditions for which they are best suited. For example, PTFE alone is too soft and has a correspondingly high wear rate. Molybdenum disulphide alone, in the best consolidated burnished film, has a very limited thickness and, therefore, a short life.

Great efforts have been made over a period of more than forty years to produce blends or mixtures which have the best possible combination of properties. The hardness and wear rate of PTFE can be improved by mixing with reinforcing fibres or fillers, and the thickness of a molybdenum disulphide film can be increased by blending it into an adhesive binder. Table 16 lists some of the many hundreds of different components which have been incorporated in solid lubricant mixtures, either experimentally or commercially. Since different combinations of any two,

Table 16 Some components of lubricating composites

Lubricant	*Matrix*	*Reinforcement*	*Other*
Molybdenum disulphide	Nylon	Glass fibre	Antimony trioxide
Tungsten diselenide	PTFE	Carbon fibre	Silica
Niobium diselenide	Acetal	Rayon	Bronze powder
PTFE	Phenolic	Molybdenum disulphide	
Graphite	Epoxy	Lead	
Silver powder	Polyimide	Silver	
Lead powder	Sodium silicate	Asbestos	
Copper powder	Bronze		
Cadmium fluoride	Silver		
Boron nitride	Cast iron		

three, four, or more of them may be used, and the proportions of each can be varied over a wide range, it is obvious that many thousands of different formulations are possible.

Technically, all such mixtures are composites, because they contain two or more solid components which are not in solution in one another. In practice, those which are used in the form of thin films tend to be referred to as solid lubricants, and those which are used as bulk solids are called composites. The most common thin film composites are probably the wide variety of bonded molybdenum disulphide films, which have been mentioned earlier. Apart from those films, the only composites which have achieved major industrial success have been those based on PTFE, which will be described in the next section.

For high temperature applications the temperature limit of PTFE is generally too low. Attempts to devise useful composites based on molybdenum disulphide have been generally unsuccessful. The most promising composite for high temperatures is a recent one based on calcium fluoride/barium fluoride.

The general philosophy in the formulation of lubricating composites is to use a continuous matrix and one or more reinforcing phases, any of which may be a solid lubricant. If neither the matrix nor the reinforcement is a solid lubricant, there will be one or more separate lubricant components. There may also be other components such as plasticizers, anti-oxidants, pigments, and even abrasives.

The temperature limits of composites are determined by those of the individual components. There is always the possibility of an adverse interaction, but the literature does not indicate that this is a serious problem. It can therefore be generally assumed that a composite can be used satisfactorily up to the limiting temperature for the least temperature-stable component. In some cases the high-temperature life of a composite may be longer than that of the least stable component, because an impervious matrix can prevent oxidation of a dispersed component, while the incorporation of anti-oxidants also extends high-temperature life.

9.6.1 PTFE composites

The most successful commercial self-lubricating composites are those based on PTFE. They are of three general types, those in which bulk PTFE is reinforced with dispersed solids, those in which PTFE fills the

Table 17 Details of some commercial PTFE composites

Type	*Filler/ reinforcement*	*Filler (%)*	*Friction*	*PV limit ($lb/in^2 \times fpm$)*
Rulon A			0.12–0.24	$15\text{–}20 \times 10^3$
Fluorocomp 103	Glass fibre	15	0.12–0.24	
Fluorocomp 174	Glass fibre	15		
	Molybdenum disulphide	5	0.14	$11\text{–}17 \times 10^3$
Fiberslip	Glass fibre		0.10–0.20	100×10^3
Glacier DU	Sintered bronze			
	Lead powder		0.10–0.20	100×10^3

interstices of a porous matrix, and those in which PTFE is reinforced by fibres. Properties of examples of all three types are shown in Table 17.

There are many examples of the first type, such as Rulon A, whose properties are shown in Table 17. Pure PTFE in block form is very soft. It will flow under high pressure, but also wears very rapidly. Filling it with dispersed solids can markedly increase its structural strength and wear resistance, but always at the expense of an increase in friction.

The wear of PTFE and its composites depends critically on maintaining a transferred film of PTFE on the counterface, and this in turn depends on the surface finish of the counterface. Against a very rough or a very smooth counterface the wear rate is high, and there is an optimum counterface surface finish. For some composites and counterface materials the optimum may be about 0.1–0.2 μm. During smooth sliding a polishing action takes place, and the surface finish of the counterface may improve beyond the optimum. The best operating conditions can actually be maintained if an abrasive filler is included in the composite.

The best example of the second type is Glacier DU, in which PTFE containing dispersed lead powder fills the pores of sintered bronze. As the composite wears, its surface becomes coated with a thin film of PTFE containing lead powder. The lead powder reduces the wear rate of the PTFE film, but does not cause as high an increase in friction as many other fillers.

There are many examples of fibre-reinforced PTFE composites, and several are listed in Table 18. Some of the best are Ampep Fiberglide in which the reinforcing fibre is cotton, and Ampep Fiberslip, where the reinforcement is glass fibre. These composites are unusual in that the reinforcing fibre and PTFE fibre are interwoven into a three-dimensional

Table 18 Some fibre-reinforced PTFE composites

Supplier	*Product*	*Composite*
Ampep	Fiberslip	Woven PTFE/glass fibre fabric phenolic bonded to backing.
	Fiberglide	Woven PTFE/rayon fabric, phenolic bonded to backing.
A.D.R.	Airflon	Woven PTFE/glass fibre.
Rose Forgrove	Unimesh	PTFE/bronze mesh.
	Uniflon R	PTFE flock-phenolic resin/Nomex fibre.
S.K.F.	Pydane	PTFE fibre/steel wire.
T.D.	Fabroid	Woven PTFE/glass fibre, phenolic bonded to backing.
	Fibriloid	Woven PTFE/Nomex fibre.
Fafnir	Faflon	PTFE/Nomex fibre.
Frankenjura	Fraslip	PTFE flock/terylene fibre.
Kahr		Terylene fibre in PTFE matrix.

weave in which the proportion of PTFE increases near the operating surface.

The Glacier DU and Ampep composites are not supplied as bulk materials, but are fabricated in the form of plain cylindrical or spherical bushes with a metal backing. These composites are capable of supporting high bearing pressures of the order of 150–400 MN/m^2 depending on the type. The operating temperature limit for the Ampep Fiberglide is determined by the cotton at about 120°C, but for the others is determined by the PTFE, namely between 250°C and 320°C. The wear life decreases rapidly towards the upper end of the temperature range.

9.6.2 Molybdenum disulphide composites

Molybdenum disulphide has been successfully incorporated in several polymer matrices, but in many cases the trade literature does not make it clear which versions contain molybdenum disulphide and which contain graphite. Nylatron GS is an example with 3 per cent of molybdenum disulphide in a nylon, and Vespel SP3 contains 15 per cent of molybdenum disulphide in a polyimide.

The 3 per cent in Nylatron GS is very low, and there must be some doubt whether at that concentration it is providing any lubrication. There is no doubt that it helps to improve the wear resistance, but this may be due to reinforcement or to some resulting change in crystallite size in processing.

Other examples have contained much higher concentrations of molybdenum disulphide in nylon, from 5 per cent to 50 per cent. At the higher concentrations the molybdenum disulphide definitely provides useful lubrication, as does the 15 per cent in Vespel SP3. Some test results have shown little reduction in friction with molybdenum disulphide concentrations up to 5 per cent, but significant reductions with 20 per cent or more.

The higher concentrations cause increasing hardness, and eventually a tendency to brittleness, so that a compromise has to be made between reduced friction and the optimum strength characteristics.

Molybdenum disulphide has often been incorporated in PTFE, as in Duroid 5813 with both molybdenum disulphide and glass fibre reinforcement. In such composites it is possible to use 10–15 per cent of molybdenum disulphide, relying on the glass fibre to maintain strength and toughness.

Incorporation of molybdenum disulphide in metals has generally been unsatisfactory, since the use of high enough concentrations to give low friction has generally caused embrittlement.

For high temperature use the best materials are therefore those with a high-temperature polymer such as polyimide, which should give satisfactory operation to over 350°C for short periods.

9.6.3 Graphite composites

Graphite has been incorporated in many polymers, but its main effect seems to be reinforcement rather than lubrication.

Cast irons might be considered as metallic composites of graphite, and the effect is to improve smooth sliding and anti-seize properties. Graphite has been incorporated in metals by powder metallurgy techniques, and the Deva metals are examples of successful commercial products. Often, however, embrittlement and loss of structural strength occur when the graphite concentration is high enough to give useful lubrication.

9.6.4 Calcium fluoride composites

A series of composites has been developed by NASA based on calcium fluoride. A major problem with them has been the fact that calcium fluoride is not effective as a lubricant below 400°C. Several attempts have been made to bridge the low-temperature gap by addition of other solid lubricants, but the difficulty has been to find an effective low-temperature lubricant which remained stable at the higher temperatures at which the calcium fluoride lubricates.

Recent work has shown that silver fulfils this requirement, and composites have been developed which provide fairly low coefficients of friction of 0.3–0.35 over the whole temperature range from 0°C to 900°C.

This composite has a matrix of 48 per cent chromium carbide bonded with 32 per cent of a nickel–cobalt alloy. The low temperature lubricant is 10 per cent of silver, and the high temperature lubricant is 10 per cent of the calcium fluoride/barium fluoride eutectic. The composite can be manufactured in bulk form by hot isostatic pressing, and is then called PM200. Alternatively, it can be plasma-sprayed onto a surface, when it is called PS200. The actual proportions of the various components can be adjusted to suit particular applications.

By lubricant standards the coefficient of friction is relatively high, and this has obvious disadvantages. If such a material were used in a low-heat-rejection engine the resulting frictional power consumption would at least partly offset the benefits of higher temperature operation. In any application the high frictional power dissipation would cause higher bearing temperatures and thus offset some of the high-temperature capability of the composite.

CHAPTER TEN

Advanced Lubrication Techniques

10.1 GENERAL

The previous chapters have shown the great difficulty which exists in providing effective lubrication at temperatures above 300°C, and the enormous benefits which will arise if the difficulty can be overcome. Considerable effort is being made to extend the limits of conventional types of lubricant and lubrication techniques to higher temperatures, and there can be no doubt that, in the long term, progress will be made.

It is clear, however, from the limited achievements of the past twenty years, that progress in developing such conventional techniques will inevitably be slow. Much thought has therefore been given in recent years to more unconventional techniques, and Table 19 lists eight possible techniques which may be useful at high temperatures.

Table 19 Unconventional lubrication techniques

Technique	Principle
(1) Gas lubrication	A gas is used in the same way as a liquid to separate bearing surfaces.
(2) Decomposing liquid	A liquid is fed to a surface at a temperature at which the liquid decomposes, but the liquid and its decomposition products give useful lubrication.
(3) Boiling liquid	A liquid is fed to a surface at a temperature at which it boils, but the liquid gives useful lubrication and cools the surface by means of its latent heat of vaporization.
(4) Liquid-transported suspension	A liquid carries a solid lubricant or the precursors for a solid lubricant in suspension. When the liquid boils or decomposes the residual solid provides lubrication.
(5) Liquid–solid–liquid	Similar to No. 4 except that the solid melts and gives higher-temperature liquid lubrication.
(6) Gas-transported suspension	Similar to No. 4 except that the carrier is a gas.
(7) Reactive gas	A gas is fed to a hot surface and reacts to give a lubricating film.
(8) Fluidized solid	A fine powder acts as a hydrodynamic lubricant, on the principle of fluidized bed technology.

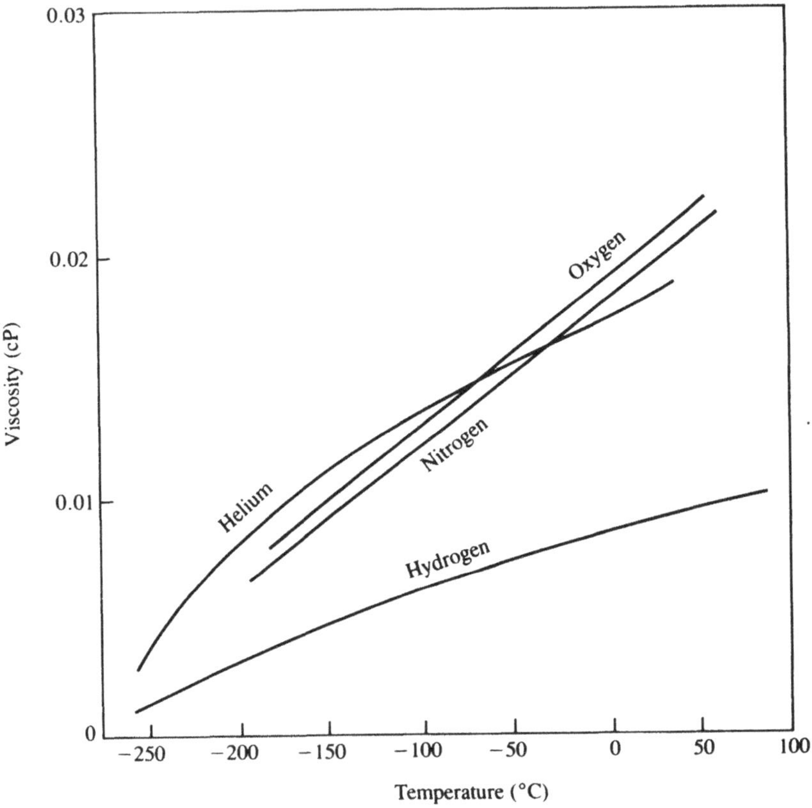

Fig. 50 Change of gas viscosity with temperature *(From* Cryogenic Fundamentals, *1971. Reproduced by permission of Academic Press)*

10.2 GAS LUBRICATION

Gas lubrication is mainly a matter of design rather than selection. It has been in widespread use for many years for specialized applications, and has some potential advantages for high temperatures because of the high thermal and oxidative stability of some of the readily-available gases.

The principles of gas lubrication are similar to those of oil lubrication, the major differences being the compressibility and the very low viscosity of gases. All gases increase in viscosity as temperature rises, and some examples are shown in Fig. 50. The resulting viscosities are still very low by the standards of lubricating oils, so that it is difficult to establish a useful lubricating film hydrodynamically, and impossible to do so elastohydrodynamically.

There are two basic ways in which gases are used. The first is analogous to the use of liquids in hydrodynamic lubrication, in that the flow of the fluid generates a film at an elevated pressure which can withstand the bearing pressures. It is known variously as gas-dynamic, self-pressurizing, or hydrodynamic gas lubrication. The second technique is to use externally pressurized gas in such a way that the applied pressure is maintained in the bearing and serves to carry the bearing loads. It is also possible to obtain a form of boundary lubrication with gases. This is not very effective, but can help to lubricate the bearing at start or stop, when the speed is not high enough for self-pressurization.

For effective gas-dynamic (self-pressurizing) lubrication very high standards of design and manufacture are required. This problem is accentuated when large temperature ranges are involved, because of the difficulty of maintaining close tolerances and clearances when considerable thermal expansion takes place, sometimes inevitably involving materials with different coefficients of thermal expansion.

Nevertheless, there are important potential advantages of gas lubrication because some of the common gases, such as nitrogen, oxygen, helium, and argon, have virtually complete chemical stability at high temperatures. The best prospect is, therefore, for externally-pressurized gas lubrication. This still requires very high standards of design and construction, but is less critical than self-pressurizing gas lubrication.

Many small-scale examples of externally-pressurized gas lubrication have been demonstrated successfully at temperatures up to 1200°C, using combustion gases and ceramics. In addition, full-sized electric motors have been operated up to 500°C with externally-pressurized air bearings, and ceramic bearings have been operated at 65 000 r/min at 800°C with externally-pressurized gas lubrication.

10.3 DECOMPOSING LIQUID

This technique is also not new. Aircraft rotary engines, which constituted a high proportion of aircraft engines in the period 1909–1920, were lubricated with castor oil on a once-through basis. This approach was necessary for two reasons. The design of the rotary engine made recirculation difficult, but in addition the rate of oxidation of castor oil at the temperature of the ring/cylinder interface was so high that re-use was not desirable.

The use of rapeseed oil in the early continuous casting of steel was a once-through technique in which complete decomposition of the oil took place rapidly at the temperature of the molten steel. Polyglycols or polyisobutenes can also be used on a single-pass basis, and have the advantage of decomposing cleanly, leaving no unacceptable residues.

In all those cases the lubrication is provided by the liquid before it decomposes, and the decomposition products make only a minor contribution. The coefficients of friction measured in laboratory tests have been in the range 0.09–0.12. This has been claimed by some workers to indicate that lubrication is by the decomposition products acting as mediocre solid lubricants. The evidence is doubtful, since starved liquid lubrication could give similar friction.

There is no doubt that some liquids produce films of degradation products, such as the cross-linked resin from silicones and the friction polymer from mineral oils. These films can certainly help to prevent metal-to-metal contact, but they have two disadvantages. The first is that neither has an inherently low friction such as that found with layer-lattice solid lubricants. The second is that it is difficult to control the rate of formation to ensure a continuous film presence without an unacceptable build-up.

A possible extension to this technique would be to use additives which would decompose to give effective solid lubrication, and this is described in section 10.5 below.

Modern environmental concerns raise two serious objections to this type of technique. The first is related to the waste of resources in single-pass lubrication techniques. The second is related to the problem of avoiding pollution from decomposing lubricant. Both objections are less relevant if the liquid decomposes cleanly and efficiently and acts as a subsidiary fuel.

10.4 BOILING LIQUID

This is a possible technique which has not yet been deliberately used with effective lubricants, but in practice a similar situation has often arisen with refrigerants or process fluids. In some types of refrigeration equipment advantage is taken of the limited lubricating ability of the liquid refrigerant, which then vaporizes as a normal part of the refrigeration cycle. Similarly, liquefied petroleum gases or other process fluids may serve both to cool and to lubricate pumps and valves before vaporizing.

10.5 LIQUID-TRANSPORTED SUSPENSION

This technique is also not new, having been used in the form of graphite-thickened polyglycol greases, pastes, or suspensions, which have been described previously. This example is effective enough for such examples as oven conveyors, and is economical, but it does not provide efficient enough lubrication for low-friction bearings or piston/cylinder interactions.

The principle could similarly be applied to the transportation of more effective solid lubricants, such as molybdenum disulphide, graphite fluoride, or calcium fluoride.

A further alternative would be to disperse the solid lubricant in a fuel or volatile recirculating liquid carrier. Such dispersions are well-known in solid lubricant technology.

An alternative possibility is to use, not a solid lubricant as such, but the precursors which, under bearing conditions, would react to produce a solid lubricant. For example, there is some evidence that molybdenum disulphide can be produced *in situ* from soluble organic molybdenum compounds decomposing in the presence of sulphur.

10.6 LIQUID–SOLID–LIQUID

This technique is only a theoretical possibility so far. An illustration of the theory is that one of the higher-melting point polyphenyl ethers might be dispersed in a fuel and fed to a combustion chamber. On evaporation or decomposition of the fuel the polyphenyl ether would melt and provide liquid lubrication.

Another illustration would be the dispersion of a low-melting point glass or boric oxide in a perfluoropolyalkylether, which would be fed to a combustion chamber or other hot system at say 700°C. The PFAE would give limited life before decomposing cleanly to leave the molten solid as a liquid lubricant from perhaps 700°C up to 1000°C. One problem would be to maintain the right amount of the second lubricant in the bearing, but this should be feasible.

This technique is similar in principle to the lubrication of some refrigerators, in which the lubricating oil is carried in dispersion in the liquid refrigerant at high pressure, and is released when the refrigerant vaporizes.

10.7 GAS-TRANSPORTED SUSPENSION

This technique is theoretically similar to the technique in section 10.5, except that the carrier is a gas. In practice, this difference is critical, since it has been found that, in a gas, initial flows soon degenerate due to agglomeration of the solid particles. The problem of maintaining stable flow of a suspension in a gas will therefore have to be solved before this technique can become practicable.

One other problem with any of these dispersion systems is that the solid lubricant would have to be very finely divided, and finely-divided solid lubricants tend to give higher friction and abrasive wear than larger particles.

10.8 REACTIVE GAS

This technique has the advantage of great simplicity compared with all the techniques using dispersions. A reactive gas can be chosen to react selectively with a specific surface, and the complete mutual solubility and diffusion of gases would ensure that the gas reaches the target surface.

Suitable reactive gases would be oxygen to produce lubricating oxides or double oxides, sulphur or hydrogen sulphide to react with molybdenum and produce molybdenum disulphide, or chlorine or hydrogen sulphide to produce EP films on steels.

The major difficulty would be to achieve the best balance between having too thin a film for effective lubrication and having too great a corrosive attack on the bearing surface. There are several methods of handling this problem. The rate of reaction can be controlled by varying the concentration of the reactive gas in an inert carrier gas, and it would in fact be possible to regulate the concentration automatically in accordance with the bearing temperature. An alternative would be to use an unreactive metal for the bearing surface, and use a reactive metal for a lubricating cage, idler gear, or other mechanical reservoir.

10.9 FLUIDIZED SOLIDS (TRIBOPARTICULATES)

One of the most interesting recent developments in lubrication technology is the investigation of the possibility of using solid particle flow to generate hydrodynamic lubrication. The technique can be considered as being analogous to fluidized bed technology, and has been pioneered by several United States workers with sponsorship of the U.S. Army Propulsion Laboratory at NASA Lewis Research Center.

Much of the work so far has involved mathematical analysis and computer simulation, but a journal bearing has been operated satisfactorily with flowing powder lubrication under a radial load of about 100 N and at DN values up to 2×10^6.

Particle sizes have been varied over a range from 0.012 μm to 10 μm, but those with good tribological properties are generally fine, with a mixture of two types of powder, some of which are free-flowing as individual grains and some of which have a tendency to agglomerate.

The general flow properties are similar to those for pseudo-plastics or Bingham plastics, in which the ratio of shear stress to shear strain decreases as the shear rate increases. They can be described by equation (57).

$$\varepsilon = \partial w/\partial y = \left(\frac{1}{G}\right)\tau + \left(\frac{1}{\eta_0}\right).\phi(\tau) \tag{57}$$

where

ε is the strain rate;
G is the shear modulus (that is interparticulate shear);
τ is the shear stress;
η_0 is the viscosity.

Effective viscosities were measured in the region of 300–400 cP for the powders studied.

Like fluids they form an intermediate layer comparable with a boundary layer on the moving surfaces, the thickness of the intermediate layer being similar in magnitude to the surface roughness.

Particle behaviour is subject to some important constraints, involving close packing and particle break-up. The study is still in its early stages, but shows useful promise for extending the scope of fluid lubrication over wider temperature ranges than the liquids currently available.

CHAPTER ELEVEN

Lubrication of Ceramics

11.1 GENERAL TRIBOLOGICAL PROPERTIES OF CERAMICS

Ceramics show great promise for use as bearing and gear materials for high-temperature applications. Structural ceramics which are of particular interest for engineering are the oxides alumina, zirconia, and chromia, together with the double oxide aluminium titanate, Al_2TiO_5, and the non-oxides silicon nitride, silicon carbide, titanium carbide, and chromium carbide.

These materials are all very hard, with high elastic modulus, and they maintain high strength to very high temperatures. They have low coefficients of thermal expansion, and low electrical and thermal conductivity. These properties are related to their high chemical bond strengths, which also lead to good resistance to oxidation and other forms of chemical attack. They are also generally considered to be very inert chemically, even to high temperatures, but in fact their chemical inertness is limited under certain circumstances, especially under rubbing conditions, and this subject is considered in more detail later.

The low coefficients of thermal expansion are usually advantageous, but may create matching problems when ceramics are used in association with metals for high temperature applications. Similarly, their low thermal conductivity is very useful in providing heat insulation, but as shown in Chapter Two, low thermal conductivity leads to high flash temperatures and high local temperatures. This results in operating surface temperatures much higher than ambient, and thus offsets to some extent the advantages of high strength and stability at high temperatures.

A major disadvantage of ceramics is brittleness, associated with the limited extent to which plastic deformation can take place. As a result there has been considerable effort to improve the toughness of ceramics, either by reinforcement or by transformation toughening, but so far most of the practical engineering ceramics are still monolithic.

The high hardness of ceramics makes them very wear-resistant in abrasive conditions. They also exhibit low wear when rubbing against softer materials such as most metals or polymers, but in these cases the correspondingly high wear of the softer materials can reduce the overall benefit.

When ceramics rub against each other under non-lubricated conditions, the improvement in wear resistance is less than would be expected from their high hardnesses. The high modulus and high hardness lead to a relatively low number of asperity contacts and high contact stresses, which cannot be relieved by plastic deformation. In conjunction with their low thermal conductivity, this leads to high local asperity temperatures, high thermal gradients, and, therefore, high thermal stresses added to the already-high mechanical stress. As a result, the overall stress pattern at asperity contacts is severe, and the benefits to wear resistance from their high hardness can be significantly reduced.

The coefficient of friction between ceramics in dry sliding is only moderate, lying generally between about 0.15 and 0.5, but it is usually steady, with little tendency to stick-slip or seizure. With even the hardest metals this is not true, due partly to the significant softening of metals produced by frictional heating.

The extent of use of ceramics in high temperature applications is growing rapidly, but in spite of their advantages their use is still very small compared with metals. This is largely due to processing cost. There are many low or moderate-temperature applications for which ceramics would be entirely suitable, but for large or complex components their manufacturing costs are several orders of magnitude higher than those of steels.

However, cost is not the only inhibiting factor. The virtual absence of plastic deformation means that stresses around manufacturing flaws or defects cannot readily be relieved. Therefore, they add to the overall stress level and can lead to fatigue failure or wear in the presence of superimposed stresses.

As a result the reliability of ceramics is not as great as for steels, so that while the use of ceramics in automotive engines has been growing rapidly, their use in critical applications such as aircraft is trivially small. An example is in their widespread use in automotive supercharger turbines, but not in the comparable aircraft gas turbines. Efforts are being made to improve quality control and reduce the incidence of manufacturing defects in ceramics, but until such improvements can be achieved, major use of ceramics will only take place where their reliability is considered acceptable, and where their advantages justify their higher cost. In the present context that means at very high temperatures.

11.2 TRIBOCHEMICAL REACTIONS OF CERAMICS

Although ceramics are considered to be chemically inert under normal conditions, they can be attacked by many reagents at high temperatures or pressures, or where the reagents are concentrated. Unfortunately, all these conditions are often met under sliding conditions, so that many tribochemical reactions can take place at ceramic rubbing surfaces.

The oxide ceramics are, of course, completely resistant to oxidation, but can be attacked by moisture or acids. Alumina will react with moisture to give aluminium hydroxide. Silicon carbide is very inert chemically, but silicon nitride will oxidize to form silica, or can react with moisture to produce silica and ammonia.

The practical result of such reactions depends on the sliding speed and load, and on the reaction rate, which in turn depends on the concentration of the reagent, the temperature, and the sliding conditions. If the reaction rate is high, then the reaction product film tends to be thicker and to break up easily. This will lead to higher wear rates and to rough surfaces. Where the reaction rate is low, thin films may be formed which will be stable and assist in lubrication.

This effect was found in one series of tests with 95 per cent alumina in which the humidity of the air was varied. At a very low relative humidity of 4 per cent the wear rate was high, but as the humidity was increased to 10 per cent and then to 50 per cent there was a slight reduction in friction and a reduction of two orders of magnitude in the wear rate. The surfaces of the worn specimens at the higher humidities were found to be partly covered with a film of aluminium hydroxide.

Under certain circumstances rubbing of reaction-bonded silicon nitride in air can develop a glassy surface layer of silica, formed by oxidation of the silicon nitride. It seems probable that under the high temperature and pressure of the rubbing contact the silica formed was liquefied and acted as a lubricant, reducing both friction and wear.

Silicon carbide is generally very inert, but in rubbing at high temperature it has been reported that silicon may be lost from the surface layers, the residual carbon being graphitized. Because the temperatures are above the range in which graphite is an effective lubricant, the decomposition results in an increase of approximately 50 per cent in the coefficient of friction, from about 0.4–0.5 to about 0.63–0.75.

A less clear picture emerged from a study in which ethylene, propane, propanol, or benzene in gaseous form was introduced during sliding of zirconia, sialon (90 per cent silicon nitride and 10 per cent alumina), or silicon carbide at temperatures from 100°C to 550°C. Coefficients of friction as low as 0.05 were apparently obtained in all cases, but the time taken to reach the lowest friction varied considerably with different combinations of gas, ceramic, and temperature, and the most rapid result was often not reached at the highest temperature. This suggests that different reaction mechanisms and possibly different products were present in different tests. The results seem very promising, but much more work would appear to be necessary before the technique could be considered for practical use. This process is an example of the reactive gas technique described in Chapter Ten.

Other chemical interactions have been found to occur with ceramics in lubricated sliding, and some of these will be discussed in the following sections of this chapter.

As with metals the problem of lubricating ceramics differs depending on whether the use is at high or low temperatures, and the two aspects will be discussed separately.

11.3 LUBRICATION OF CERAMICS AT LOWER TEMPERATURES

For the purposes of this discussion, low temperatures will be taken to mean temperatures low enough for conventional lubricants and additive technology to be used, probably below about 250°C.

The current consensus of opinion seems to be that conventional hydrodynamic lubrication can be satisfactorily used for ceramics, provided that the usual standards of surface finish and tolerances can be maintained. There are many design constraints which have to be taken into account, such as their generally lower thermal expansion and low thermal conductivity, and the need to avoid high tensile stresses and stress concentrations.

Where full fluid film lubrication cannot be maintained, the situation becomes more complicated. Esters and mineral oils without additives have been found to be relatively ineffective in boundary conditions with silicon nitride, whereas a hydroxy-terminated polyglycol gave better performance. This is in contrast to steels, where esters are effective in boundary lubrication. Under boundary conditions both silicones and perfluoroalkyl polyethers are believed to strip oxygen from surfaces, so that these fluids are not likely to be usable with oxide ceramics, even at low temperature.

It has always been obvious that existing boundary additives, developed and optimized for metal surfaces, were unlikely to be optimum for the completely different chemical environment of ceramic surfaces. It was even probable that any chemistry designed for oxide type ceramics would not be ideal for nitrides, carbides, or silicides.

So far there has been relatively little investigation of the effect of the present generation of boundary additives on ceramics, but there has been enough to show differences in behaviour. One of the earliest studies showed that at 90°C two different ZDDPs, which are universally effective anti-wear additives for steel, were almost completely ineffective with alumina, but effective for silicon nitride, silicon carbide, and partially stabilized zirconia (PSZ). Similarly, lauric acid, which is a mild anti-wear additive for steel, alumina and PSZ, is corrosive to silicon nitride under rubbing conditions. At lower speed and temperature, where the dominant mechanism is adsorption, the differences are not found, and it is clear that some rise in temperature is necessary for the reactions to occur.

The corrosive effect of lauric acid on silicon nitride is a matter of some concern, since organic acids of similar type, and often more reactive, are produced as oxidation products in automotive engine oils. A similar test at lower temperature with oleic acid on silicon nitride was reported to give low friction and wear, so it is clear that further work is desirable to clarify this behaviour with long-chain aliphatic acids.

More recent work with other phosphorus compounds on silicon nitride has also shown them to reduce friction and wear in boundary conditions, but no comparable tests were reported with other ceramics.

In research aimed at raising the operating temperature of automotive engines, several different ceramics have been used with conventional additive-containing mineral oils, synthetic oils, or esters. No particular boundary lubrication or wear problems have been reported, although the ceramics have included chromium oxide, alumina, chromium carbide, and titanium carbide.

In summary, it could be said that no great difficulty has been found in lubricating ceramics at low temperatures, but that there are several indications that problems could arise. More useful information on the matching of base oil, additives, and the type of ceramic, is likely to become available as more experience in the use of engineering ceramics is obtained. In particular, there is, as yet, very little information about performance above 100°C.

11.4 HIGH-TEMPERATURE LUBRICATION OF CERAMICS

The problems of long-term lubrication at temperatures above 260°C have, of course, not been solved even for steels or refractory metals. The amount of investigation into this area for ceramics so far is very much less. The major collaborative study programmes which have been set up in various countries, such as the British CARE programme, have not yet published much information from their work, but the indications are that the greatest effort has gone into mechanical and manufacturing studies, and not into lubrication.

The most promising class of oils for long-term use above 300°C are the perfluoropolyalkyl ethers, but they do not yet seem to have been tested at high temperatures with ceramics. The oxide-stripping from metals which was mentioned in Chapter Seven would appear to eliminate them from consideration for high-temperature use with the oxides, which include some of the most promising ceramics.

The alternative approach for lubrication of ceramics at high temperatures is the use of solid lubricants or coatings. At intermediate temperatures, to perhaps 400°C, the most promising candidate solid lubricant is molybdenum disulphide. Many different techniques have been studied for applying molybdenum disulphide coatings to various ceramics, but this has proved much more difficult than with metal surfaces.

The simplest technique for coating on steel is burnishing, in which the molybdenum disulphide powder is rubbed onto the surface by means of some form of flexible pad. This is much less successful with ceramics than with metals, probably because the high hardness of ceramics prevents the slight embedding of the crystallites which is believed to be necessary for the initial attachment of a burnished film.

The use of bonded coatings is also more difficult. Even on metals some pre-treatment is necessary for the achievement of strong bonding, such as grit-blasting to provide an optimum surface roughness, and a chemical conversion coating such as phosphating of steels or anodizing of aluminium. So far there has been insufficient investigation to establish suitable pre-treatments for ceramics. A further difficulty is that most of the available bonding materials are organic and would be unsuitable for temperatures over 250–300°C.

Sputtering cannot be achieved by conventional DC or RF (radio-frequency) sputtering techniques because the lack of electrical conductivity in the ceramics leads to generation of a DC potential on the surface which repels the sputtering ions. Recent work has succeeded in sputtering silver onto ceramics by using a silver screen to prevent DC potential. The sputtered silver is a fair lubricant in itself, but may also assist the sputtering of other substances by preventing DC potential. Some limited success has been achieved with sputtering of molybdenum disulphide onto silicon nitride, but no reliable standard technique has yet been developed.

Other techniques which have been studied include ion plating, plasma deposition, and ion beam enhanced deposition. In some of them a mixture of molybdenum disulphide with nickel, gold, chromium silicide, or boron carbide has been applied. Promising results have been achieved at a research level, and there has been some practical application in spacecraft, but there is no indication that any of the processes is suitable for less exotic commercial applications. One important factor is that the friction of molybdenum disulphide is lower (down to 0.01) and the upper temperature limit is higher (more than 900°C) in space vacuum, so that lower wear rates and longer life can be achieved than in normal oxygen-containing atmospheres.

On an experimental basis some promising results have also been reported in the lubrication of ceramics with graphite, but these are also subject to the same difficulties in establishing reliable coating techniques. With adsorbed nickel chloride the useful range of the graphite was increased so that it could be used satisfactorily over most temperatures from ambient to its oxidation limit at about 500–550°C, with a coefficient of friction rising from 0.17 to 0.25. The intercalated graphite in this test series was applied by transfer from an adjacent rubbing pin applied to the same test disc.

For higher-temperature applications in normal atmospheres, coatings such as the calcium fluoride/barium fluoride composite PS200 described in Chapter Nine appear to have the greatest promise. PS200 has in fact been successfully plasma-sprayed onto partially-stabilized zirconia. However, the coefficient of friction of PS200 is about 0.3. Hexagonal boron nitride has also been applied successfully by a transfer technique, and has a coefficient of friction between 0.2 and 0.3, but it can only be used at temperatures above 300°C.

An interesting and unconventional technique has been used to generate a film of tricresyl phosphate (TCP) on a ceramic bearing surface. The liquid TCP was vaporized at 300°C into a high purity nitrogen carrier gas, to give a concentration in the gas of 1.5 per cent by volume. The gas was then fed into an oven at 700°C containing a specimen of hot-pressed silicon carbide or silicon nitride. The result was the formation on the target specimen of a film which was considered to consist of partly-degraded TCP, possibly slightly polymerized.

Satisfactory films were produced on metals by this technique without pre-treatment, but in order to produce a tenacious film on a ceramic, it was necessary to pre-treat the surface with a solution of ferric chloride or ferric acetylacetonate. This suggests that some similar form of pre-treatment might usefully be studied for improving the adhesion of other coatings, such as burnished or bonded solid lubricants.

Overall, it can be seen that ceramics can be lubricated satisfactorily at low temperatures, provided that no new problems of chemical incompatibility arise. At higher temperatures the techniques available are limited and largely experimental. Above the highest temperatures at which molybdenum disulphide can be used, probably about 400°C in air, coefficients of friction greater than 0.2 are probably unavoidable with existing technologies.

CHAPTER TWELVE

Automotive Lubrication

12.1 GENERAL

The lubrication of automotive engines occupies the attention of more specialists than any other field of lubrication. This chapter is not intended for those specialists, since they know far more about the subject than anyone not active in the field can possibly do. This is not simply a problem of the vast amount of published and patent literature which exists on the subject. Automotive lubrication and engine design are two of the most highly competitive fields of commercial research and development. The latest developments are usually kept confidential until either they are approaching the stage of commercial operations or they are considered to offer no competitive advantage.

There are two reasons for including a chapter on automotive engine lubrication here. The first is that, for those who are not lubrication specialists, the lubrication of motor vehicles is usually the most interesting aspect of lubrication, because a motor vehicle is generally the second most expensive item in a person's life, after their home.

The second reason is the recent surge of interest in the development of low-heat-rejection, or 'adiabatic' engines. This interest is not restricted to automotive engines, but the great commercial and environmental pressure is for automotive development, because the automotive engine is the greatest single user of fossil fuels and the greatest single generator of atmospheric carbon dioxide.

There are, of course, many lubricated components in a motor vehicle apart from the engine. They include manual or automatic gearboxes, rear axle/differentials, wheel bearings, brake systems, steering systems, shock absorbers, and numerous small pivots, hinges, latches, and so on.

Lubrication problems in motor vehicles are relatively uncommon. When they do arise they tend to fall into one of three categories.

(1) Failure by the user or the servicing organization to follow the recommended lubrication practices.
(2) Operation of the vehicle in some way which is at or beyond the limits for which it was designed.

(3) Innovative design features, or extrapolation of a conventional design to higher loads, speeds, or powers.

The second and third of these categories often lead to high temperatures, but alternatively may lead to excessive wear, seizure, or fracture of a component. Examples of the second category were quoted in Chapter Two. The third category sometimes involves many thousands of vehicles and causes huge expense and notoriety. Some examples are the scuffing of diesel engines, wear of cams and cam followers, and the 'mayonnaise' emulsion which sometimes forms in engines, and which has occasionally reached plague proportions.

The case of scuffing will be described in some detail because it is an important aspect of failure of lubrication associated with high temperatures.

Excessive temperatures may occur in gearboxes, rear axles, or wheel bearings, especially when a vehicle is operated in an exceptionally high-temperature environment, such as in long journeys over desert roads. Excessive temperatures in brakes are usually caused by over-use, especially in hot environments, and are more common in drum brakes than in disc brakes because of a relative lack of cooling airflow.

For the user, the best preventive action is to reduce the amount of continuous use, and allow periods for cooling. It this is not possible, or if it is known that a vehicle or component has been subjected to unusually high temperatures, it is important to remember that the lubricant may be subject to more rapid deterioration. It may, therefore, be beneficial if inspection or re-lubrication can be carried out more frequently than usual.

12.2 CONVENTIONAL AUTOMOTIVE ENGINES

Most automotive engines are liquid-cooled upright in-line four-stroke reciprocating engines with four, six, eight, or twelve cylinders in one or two banks. The most common exceptions are the smaller two or four-cylinder air-cooled engines, often two-stroke, widely used in motor cycles, but also used in some smaller cars such as the Volkswagen 'Beetle'. Other types of engine, such as the Wankel and gas turbines, have been very little used for road vehicles.

Numerically most automotive engines are spark-ignition (petrol or gasoline) engines. Until a few years ago the compression-ignition (diesel) engines were almost entirely limited to the larger commercial vehicles, but recently diesels have increasingly penetrated the light commercial and private car markets. From the point of view of lubrication temperatures there is little difference between spark-ignition and diesel engines. Both are designed to use the highest temperatures which the engine materials and lubricants can withstand, so that even with many different design features, the operating conditions are determined by the temperature limits, rather than vice versa.

Internal combustion piston engines are an interesting illustration of variations in temperature in a machine and the corresponding times of exposure of the lubricants to those temperatures. Combustion gas temperatures are typically between 1600°C and 2000°C, and the temperature of the cylinder wall above the top piston ring may reach 1000°C. No lubrication is needed, and although a small quantity of lubricant will reach these areas, it would be rapidly degraded.

The hottest part needing effective lubrication is the top piston ring, where the lubricant may reach 200–260°C. The volume of oil in the ring groove and at the ring/cylinder interface at any one time may be less than 1 ml and its total time of exposure will be in seconds rather than minutes. The extent of oxidation and thermal degradation in that small volume may be high, but its contribution to degradation of the whole oil fill will be relatively small.

There is evidence that 40–70 per cent of the base oil will evaporate from the oil in the top ring groove. This evaporation will help to reduce the peak temperature of the remaining oil, but will also have the effect of leaving higher viscosity and a much higher additive concentration in the residual oil, so that both boundary lubrication and oxidation resistance will be improved.

The evaporation of the base oil is not only due to the volatility of the lower-molecular-weight components, but is almost certainly also caused by thermal cracking or oxidation of the oil molecules to produce smaller molecules. Apart from loss of base oil, analyses have also shown that the oil and additives are quite badly degraded.

The piston skirt may be at 140–170°C, but the small-end bearing (gudgeon pin bush) is cooled by oil passing through the connecting rod and oil spray in the crankcase, so that its lubricant film is likely to be at 120–150°C. There is a high oil flow-rate in this region, and the oil as a whole may spend 5 per cent of its life at these temperatures.

The oil in the sump may eventually stabilize at between 100°C and

140°C and the oil film in the main and big-end bearings, at 100–150°C, and these are the temperatures which dominate the oxidation of the oil.

There have been dramatic improvements in the life of engine oils in recent years. In a typical passenger car the oil change period was 1500 km in 1949, but this increased to 10 000 km by 1972 and 15 000 km by 1992. These increases are in spite of a concurrent increase in operating temperatures.

The consensus of opinion among automotive lubricant specialists is that no quantum leap is likely in automotive engines in the near future. There is a steady improvement in terms of fuel consumption and specific power, with resulting increases in operating temperatures. Sump temperatures in particular have risen during the past forty years by an average of perhaps 30°C, but in the same period oil change periods have increased ten-fold.

The general trend to higher specific power leads to higher operating temperatures, but there are also several environmental concerns which are influencing automotive lubricant development.

It has already been mentioned that combustion efficiency improves as the combustion temperature increases, so that the pollution associated with unburned and partly-burned fuel is decreased if the combustion temperature rises. A major step forward in this respect is the so-called 'lean-burn' engine which is designed to use fuel/air ratios much weaker than present engines. Lean-burn engines should generate less exhaust contaminants and particulate emission, but at the expense of some increase in combustion temperature. However, the problems in producing a lean-burn engine with stable combustion have not yet been entirely solved.

A significant part of oil consumption is due to oil volatilization in the combustion zone, so that there is considerable pressure to reduce oil volatility. This may have some adverse effect on performance, since, as pointed out previously, oil vaporization has beneficial effects in reducing the peak oil temperature at the top ring and increasing the concentration of additives in the residual oil.

The desire to reduce oil volatility is also in conflict with the target of reducing frictional power loss and, therefore, fuel consumption, by the use of lower-viscosity oils. Low-viscosity oil usually means low molecular-weight, which in turn means higher volatility, so that the use of mineral oils is being squeezed between the requirements of low viscosity and low volatility. This pressure will probably result in more rapid introduction of synthetic hydrocarbons and esters, which are better able to meet both requirements. Some synthetic hydrocarbons do not give sufficient swell to seal rubbers, and the addition of a small quantity of an ester to a synthetic hydrocarbon can provide the correct swell, so that it may be generally beneficial to use blends of synthetic hydrocarbon and ester.

Synthetic oils, whether hydrocarbons or polyol esters, have better viscosity–temperature characteristics than mineral oils and therefore require lower concentrations of viscosity-index improvers. This is a mixed blessing, because the temporary shear loss with VI improvers in high-speed contacts helps to reduce viscosity and, as a result, viscous friction and power loss. Higher operating temperatures are, in fact, leading to problems of low oil viscosity, and the temperature limit for an oil may be imposed by viscosity index rather than by thermal or oxidative stability.

Overall, European vehicles tend to run with higher sump temperatures than American vehicles, partly because of their higher compression ratios, although in both regions there is considerable variation between north and south. At present there is greater pressure in North America for low volatility and oil purity, and less for high viscosity index, so that a higher proportion of synthetic hydrocarbons is being used, but also significant amounts of severely-hydrogenated oils, which have low viscosity indices.

In Europe the utilization of synthetic base stocks is still very low, but it is likely to grow rapidly. The need for high viscosity indices will lead to greater use of PAOs and heavily hydrocracked base stocks.

One other environmental pressure which has not been mentioned is the requirement for a high level of biodegradability. This will encourage the use of the long-chain polyol esters, but has also recently led to the introduction of a new vegetable-oil based automotive engine oil, for the first time in over fifty years.

12.3 SCUFFING OF CYLINDER LINERS

Scuffing is a form of severe adhesive wear which results in serious surface damage and often in seizure or component fracture. It is known in the United States as scoring. It can occur in many types of component, including gears, plain bearings, cams, and tappets, but this discussion will be restricted to the subject of scuffing in automotive piston rings and cylinder liners. This is a recurring problem whose occurrence and severity vary from time to time, but which has never completely disappeared.

Scuffed surfaces have a welded and torn appearance, often with a directional trend parallel to the cylinder axis. The material gives the impression of having been softened, or even liquefied at the surface, and the surface of the damaged area is generally smooth rather than roughened or jagged. Metallographically the damaged region is found to contain areas of a very hard white, non-etching, transformation alloy, and in some instances this hard material has been shown to consist of ledeburite. Ledeburite is a very hard, high-carbon eutectic ferrous alloy which requires temperatures over 1100°C for its formation.

There is no complete agreement about the mechanism of scuffing. In piston ring/cylinder interactions scuffing always begins at or near the top dead centre position of the top ring. This is the point at which the motion of the ring relative to the cylinder ceases momentarily, and there is no axial movement to create a hydrodynamic oil film. The residual oil film is therefore dependent on inertia and on the viscosity of the oil, very low at that point, resisting any squeeze film effect. This is the point at which the thickness of the oil film is a minimum, and some measurements have indicated that the oil film thickness reduces to zero, resulting in mixed or boundary lubrication.

Thermal equilibrium at any lubricated contact depends on all heat input, including frictional heat, being balanced by heat removal. At the top ring at top dead centre there are significant heat inputs from the combustion, and heat removal is mainly by the engine cooling system and partly by the lubricating oil.

It seems probable that scuffing begins when the thermal equilibrium is lost instantaneously, possibly because of a transient overload or a solid particle or a local interruption in the oil supply causing a reduction in the oil film thickness. This would have the effect of converting full oil film separation of the surfaces at that point into mixed or even full boundary lubrication, with a resulting increase in friction.

The local increase in frictional energy dissipation will further increase the temperature, and this could have two adverse effects. If a local hot spot develops, then expansion of the metal will produce a high spot, which will further reduce the oil film thickness. Any increase in temperature will also reduce the oil viscosity, further reducing the oil film thickness.

Unless the initial transient initiating event is removed in time for the increased heat to be removed, the system may become unstable. The increased temperature leads to reduced oil film thickness, which causes higher friction and further heating. The metal temperature at this point is likely initially to be greater than 350°C. Once the oil film becomes too thin to prevent asperity contact, flash temperatures will increase the temperatures at the contacts, and local asperity welding will become probable, followed by breaking of the welds.

This will in turn cause further heat build-up, and ultimately a point must be reached where larger areas of the two surfaces become softened and contact one another, leading to very high local friction, and area welding and tearing, or, in other words, scuffing.

There is some debate about whether the presence of a lubricant is required for scuffing to occur. The composition of the ledeburite phase which has been found in scuffed areas requires more carbon than is present in steels, and this additional carbon is presumably derived from the lubricant. With cast iron, however, the free carbon present would be sufficient for the formation of ledeburite.

Prevention of scuffing is basically by ensuring the presence of an adequate oil film thickness. There is usually little scope for achieving this by the use of a higher-viscosity oil, but development of a suitable curved profile on the ring is very helpful. This is usually achieved during running-in, so it is important for engines to be properly run in, and this is now commonly assisted by the use of a specially-formulated oil during the first few hours of operation. Another useful technique is the production of a suitable surface texture on the surface of the cylinder liner. Very smooth surface finishes tend to increase the incidence of scuffing.

Scuffing is best avoided by using pairs of materials for piston ring and cylinder liner which do not readily adhere or weld together. This implies avoiding the use of two ferrous alloys. As engine temperatures increase in the search for improved fuel efficiency, the risk of scuffing is also likely to increase, and ceramic coatings have been used on top rings and cylinder liners in the experimental low-heat-rejection engines which are described in the next section.

12.4 LOW-HEAT-REJECTION ENGINES

Conventional internal combustion engines have very low fuel efficiencies, generally between 20 per cent and 30 per cent. Much of the fuel wastage is due to the need to cool the engine, which involves direct power consumption for cooling and lower combustion efficiency. A significant improvement in fuel efficiency could be achieved if engine temperatures could be allowed to rise by eliminating artificial cooling.

The increased fuel efficiency would have two important effects, a major reduction in consumption of fossil fuels, and a major reduction in the generation of atmospheric carbon dioxide. The great current interest in higher-temperature engines was initiated by the rapid increase in crude oil costs in the 1970s, and the resulting concern about the remaining life of the world's petroleum reserves. That concern has abated to some extent at present, but has been replaced by deepening concern about the 'greenhouse effect' caused by high atmospheric carbon dioxide.

The new high-temperature target engines were originally referred to as adiabatic engines. This is incorrect because the word adiabatic means operation without heat entering or leaving the system. This is obviously inappropriate for an internal combustion engine, and the preferred title is 'low-heat-rejection' (LHR) engines, or alternatively 'high-temperature' or 'insulated' engines.

One of the main reasons for the level of cooling in conventional engines is the temperature limitations on the use of the lubricating oil. As has been shown earlier in this book, the limiting temperature for a particular oil is not absolute, but depends on the time of exposure in relation to the required oil life, as well as the acceptable level of oil deterioration.

Obviously, therefore, there is scope for some increase in temperature even with present oils, provided this is coupled with a reduction in the oil-change interval. The latter is considered commercially undesirable, and is contrary to the general trend in recent years. The resulting small increase in operating temperature would have some beneficial effect on fuel efficiency.

However, the major stepwise improvement would come with the elimination of all artificial cooling systems, and this would inevitably cause a major increase in temperature. The actual temperatures involved would still be subject to various design decisions, such as cylinder number, size, and layout, compression ratio, and speed, and the even more basic decisions as to whether the engine is a turbine, rotary piston, Stirling, conventional diesel or spark ignition, two-stroke or four-stroke .

As a result there has been considerable discussion about the best target engine, and a variety of different sets of temperature predictions has

Table 20 Typical temperature predictions for a low-heat-rejection diesel engine

Component	*Predicted temperature (°C)*
Rings and cylinder liners	320–595
Valves and guides	430–650
Tappets	370–430
Cams	180–205
Rolling bearings	260–650
Plain bearings	260–650
Splines	540–650
Gears	150–320

appeared. A typical set of target temperatures is given in Table 20 for an LHR diesel engine.

It can be seen from the table that the design envisaged is fairly conventional, and this seems to be characteristic of studies of LHR engines. For example, a DTI brochure about the British CARE (Ceramic Applications in Reciprocating Engines) consortium's fields of interest makes it clear that all the activities assumed a design very similar to that of current conventional engines. This situation is perhaps understandable in the early stages of LHR engine development, because of the mass of experience which exists in conventional automotive engine design, but in the long term it seems extremely unlikely that a format optimised for low temperatures will also be best for much higher temperatures.

Probably the most successful projects to date have been sponsored by the U.S. Army Tank and Automotive Command (TACOM). There are two basic projects, based on a Cummins Big Cam NT turbo-compound diesel engine and on a 230 hp LHR engine installed in a U.S. Army 5-ton truck. Both engines were basically standard production engines, with minor components modified to eliminate problem areas.

Ceramic coatings have been used on piston rings and cylinder liners, the most recent being chromium carbide on the top rings and chromium oxide coatings on the cylinder liners. The truck installation has easily completed a standard NATO 400 hour engine test cycle with an oil change period of 150 hours using a pentaerythritol (polyol) ester lubricant specifically developed for this project. Since then more advanced lubricants based on higher-viscosity polyol esters or polyphenyl ethers have been developed and tested.

Similar work in Japan on a heat-insulated 6-cylinder turbo-compound diesel engine used a fully-formulated mineral oil, but polyol esters have also been studied as better potential lubricants for such engines. In this engine the top rings were plasma-sprayed titanium carbide, and the liners plasma sprayed chromium carbide.

In the United Kingdom, Leyland DAF have studied the effect of reduced water cooling of the cylinder head and introduced a number of ceramic components. Those changes permitted a reduction in engine size and reduced cooling fan power, resulting in an overall 10–12 per cent improvement in fuel consumption. Thermal analysis suggested a probable top ring groove temperature of 300–350°C, and it was found necessary to use a part synthetic mineral oil blend.

It is not clear what the equilibrium temperature of the sump oil has been in these studies. One important effect of the elimination of cooling is to raise the temperature of the oil returning to the sump, and a reasonable assumption is of a sump oil temperature of 175°C, compared with perhaps 100–140°C for typical present engines.

Reference to any of the temperature limit figures in Chapter Seven shows that this is likely to shorten the oil life by a factor of 25–30. Polyol esters are seen to have a useful life at a sump temperature of 175°C, but any further increases in sump temperature for more advanced LHR engines may well force a reduction in oil change periods, even with polyol esters. The use of polyphenyl ethers would require pre-heating to ensure engine starting, and PFAE cannot yet be considered sufficiently well developed.

It must be emphasized that these project engines have all represented an interim stage in the development of true LHR engines, but all of them have demonstrated reductions in fuel consumption with acceptable oil performance. It is less clear what the total engine life would be, since the U.S. and Japanese engines have both needed wear-resistant coatings in order to prevent serious wear. The life of those wear-resistant coatings is likely to be much less than the lives of well over 300 000 km, say 8000 hr, achieved by present diesel engines.

The Japanese manufacturer Isuzu is apparently still planning to market a non-cooled automotive engine in 1995, but the actual level of insulation may well be the major unknown factor in ensuring the commercial feasibility of the project.

It is difficult to obtain information about the latest progress in LHR engine development or lubrication at present, because both engine and lubricant manufacturers are keeping their work confidential.

12.5 OTHER AUTOMOTIVE LUBRICANTS

One striking development in chassis lubrication during the last thirty years has been the elimination of the need for re-greasing of most suspension joints in cars and other light vehicles. This has been achieved by the use of high quality greases, often containing some molybdenum disulphide.

Lubricants in some other vehicle components are also often expected to last the life of the vehicle, including rear axles, automatic transmissions, and gearboxes, where these are separately lubricated from the engine. Some of these components are seeing increasingly high temperatures, and there is a slow move towards the introduction of synthetic hydrocarbons and esters in such items.

Until recently brake fluids were almost universally polyglycols, although castor oil, mineral oils, and silicones had previously been used. In a sealed system with limited oxygen contact, polyglycols are able to cope with transient temperatures up to 300°C in overheated brakes, and because of the 'unzipping' breakdown mechanism, the degradation products are low-molecular-weight compounds which remain in solution in the system until they are able to evaporate in the reservoir. As a result there is little risk of vapour-locking or pipe blockage, but only a loss of fluid volume which needs replacement.

The main problem with the polyglycols is water absorption, and increased standards in this respect have led to widespread use of borate esters. These react with water and thus remove it from the system, but they are used as mixtures with polyglycol to maintain the required rubber swell, pour points, and boundary lubrication. Different reports have referred to mixtures containing anything from 30 per cent to 90 per cent of borate ester, the balance being polyglycol.

CHAPTER THIRTEEN

Selection of Lubricants

13.1 GENERAL

It is a curious fact that as the required temperature increases, the problem of lubricant selection becomes simpler. On the basis that 'Man who has choice has problems', selection for applications at 100°C is very difficult, while selection for temperatures above 1000°C is very simple: there is virtually no choice.

The selection problem changes progressively as the design temperature rises. An ideal approach to the overall selection process would be a computer-based procedure in which all the design parameters, including the various temperature profiles, could be entered. It is very doubtful if the present state of the art of lubrication would permit such an approach. An attempt is currently being made in the United States to construct a data-base on which such a selection process for normal temperatures could be based, and it will be interesting, in due course, to see the degree of success which is achieved.

Lubricant selection should always be part of an iterative design process, so that the effect of each possible lubricant choice on the other aspects of design could be progressively assessed as the design proceeds. In practice, such a situation is rare or non-existent. More often a design approaches finalization before any real thought is given to lubrication.

It follows that, in most real situations, the choice of lubricant is determined by the nature of the finalized design, rather than by any value judgment about the merits of different lubricant choices in the broad conditions of the application.

For the lubricant specialist, guidance on lubricant selection has therefore traditionally taken the form of defining the limits of applicability of the various classes of lubricant, together with advising on the general criteria which determine the choice of class. Once the choice between oil, grease, gas, or solid lubricant has been made, or happened by default, any well-trained engineer can carry out the detailed design or selection of the bearing, gear, or other component and determine the viscosity grade of the oil or the NLGI grade of the grease, and so on.

The problem of selection in this chapter will therefore be approached in the same way, by defining the limits of applicability of the various lubricant classes in a series of temperature ranges.

13.2 NORMAL TEMPERATURES; −40°C TO 200°C

For much of this range, a very wide variety of lubricant types and brands is available, and the choice will very often be made on a price basis. There are also many different guides available to lubricant selection. Figure 51 is taken from *The Tribology Handbook*, and shows the approximate speed and load limits for the use of oils, greases, and solid lubricants. It can be seen that, at tangential speeds between 20 and 250 m/s, only oil can be used, although gas lubrication is usable over the whole speed range.

Below 0.6 m/s and 5 MN/m^2 any of the four classes of lubricant can be used, and the designer can choose the type which offers the most advantages.

It is difficult to portray graphically the influence of temperature on this selection. Below 90°C temperature has little effect on selection. Above 90°C Fig. 20 shows that the life of mineral oils is diminished, and it has been said that a device which requires a very long lubricant life, such as an automotive gearbox, should use a synthetic hydrocarbon oil if the oil temperature is consistently above 90°C. Phosphate esters are approaching their limit for long-term operation at this temperature.

Above 90°C some of the simpler greases may be inadequate, especially in some small devices such as miniature electric motors, where the limiting temperatures may be as much as 20–30°C below those applying in larger systems. As the temperature approaches 200°C all the simple soap-thickened greases become unusable.

As the temperature rises from 100°C to 200°C all the conventional oils and greases, including mineral, synthetic hydrocarbon, di-ester, polyol ester, polyglycol, silicate ester, and silicone will have life limitations.

As far as solid lubricants are concerned these temperatures are well within the capabilities of PTFE, graphite, and molybdenum disulphide. It should be remembered, however, that the speed limit of 0.16 m/s for dry lubricants in Fig. 51 is imposed by the problem of heat dissipation, so that if the ambient temperature rises, the limiting speed may have to be reduced. This also applies to greases.

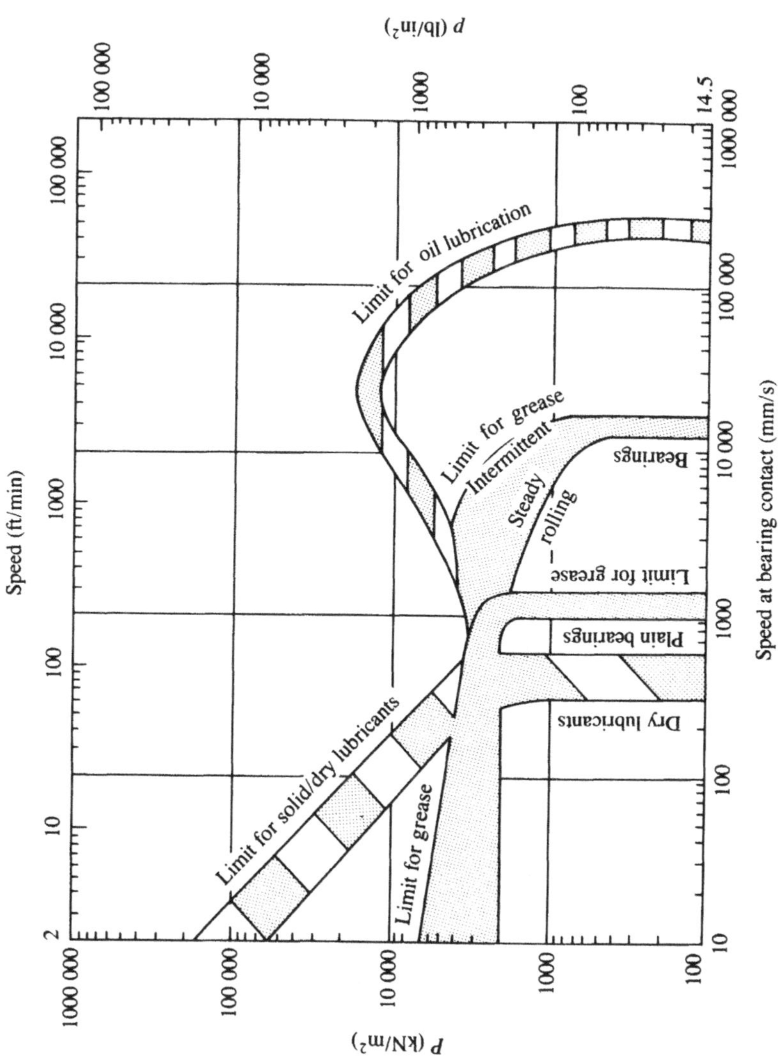

Fig. 51 Speed/load limitations for different types of lubricant

13.3 THE TEMPERATURE RANGE 200°C TO 260°C

Within this range many of the conventional oils and greases can still be used, but their life will be relatively short. For example, cylinder head temperatures in many aircraft piston engines have been well into this range for many years, but mineral oils have been satisfactory because the period of exposure to upper cylinder temperatures is very short.

For long lives in this range, only the more stable synthetic oils such as polyol esters and phenyl silicones can be used, and even they become seriously life-limited towards the top end of the range. This is the range in which polyphenyl ethers and perfluoropolyalkyl ethers become serious contenders for some applications.

This is also the temperature range in which the 'clean-burning' oils, such as polyglycols and the polybutene synthetic hydrocarbons, begin to be useful, since if decomposition is inevitable it is obviously desirable to minimize the risk of undesirable decomposition products.

The higher-temperature complex greases and some of the non-soap greases are usable within this range, but the tangential speed limits become progressively lower as the temperature rises because of the additional thermal stress caused by frictional heating. The simple soap-thickened greases are unusable at these temperatures.

PTFE is unusable in this range except for short periods or low loads with the highly-reinforced composites. Graphite can be used in this range, but its friction and wear rate will be high, except in very moist conditions (e.g., in steam) or when some less volatile adsorbed substance is added to it.

Most of the commoner polymers also become unusable as binders or components of composites, but polyimides, polyphenylene sulphides, and silicone resins are still usable.

Molybdenum disulphide is still well within its usable range, but some of the commoner bonded films which are epoxy or phenolic-bonded will deteriorate. The choice is then limited to the three polymers mentioned in the previous paragraph, or to inorganic binders, or burnished or sputtered films.

13.4 THE TEMPERATURE RANGE 260°C TO 350°C

The only oils which are usable for any length of time in this range are the polyphenyl ethers and perfluoropolyalkyl ethers, and they become life-limited at the upper end. The life of all other oils is so short at these temperatures that they are best used in single-pass, or non-recovered, applications. Under those circumstances there are advantages in using those oils which give the least undesirable combustion products, including synthetic hydrocarbons, rapeseed oil, and polyglycols.

This is the temperature range in which graphite-thickened polyglycols are useful, because the polyglycol provides fluidity for feeding to the lubrication point, then decomposes cleanly to leave graphite, which is again becoming a useful lubricant at the top end of the range. Graphite survives long enough to provide useful lubrication and then decomposes in turn to leave little or no deposit.

Molybdenum disulphide is still useful in this range. Some of the other inorganic solid lubricants – lead monoxide, lead sulphide, molybdic oxide, and boron nitride – become progressively better lubricants as the temperature rises through the range.

The only grease which has a useful life in this temperature range is the PTFE-thickened perfluoropolyalkyl ether.

13.5 TEMPERATURES HIGHER THAN 350°C

At temperatures above 350°C the only liquids with a long life are the liquid metals, liquid oxides, and glasses. It is still possible to use some of the organic liquids, but their lives are very short. Even so, oils such as polyglycols, synthetic hydrocarbons, and rapeseed oil can be used for very short periods at these temperatures, and rapeseed oil has been used in contact with liquid steel at 1200°C in continuous casting. Its life in contact with the molten steel at such temperatures is certainly less than seconds, but its latent heat of vaporization and its low thermal conductivity enable the bulk oil to remain present in the vicinity of the molten steel for many seconds. In this application the major advantage of the rapeseed oil is that it does not cause carburization of the surface of the steel strand.

Molybdenum disulphide has only a reduced life in air above 350°C, but can be used to over 900°C in vacuum and probably in certain types of inert atmosphere. Molybdenum disulphide and some of the other lubricating dichalcogenides are in fact the only efficient lubricants which are usable from very low temperatures to such very high temperatures. They suffer from the usual disadvantages of solid lubricants, as well as from their increasingly poor oxidative stability as the temperature rises above 350°C.

Where there is a need to achieve a coefficient of friction lower than 0.1 at high temperatures, it may be necessary to restate the lubrication problem. Instead of asking what we can use to lubricate a system at a specific high temperature, we have to start asking how we can alter the system in order to use molybdenum disulphide satisfactorily at that temperature.

There are other possible lubricants for use up to 900°C, of course, even in air. The variety of lubricating composites has been mentioned in Chapter Nine, and some of the non-layer-lattice inorganic solid lubricants can be used over most of the range from 350°C to 900°C or 1000°C. Probably the best so far is the calcium fluoride/barium fluoride composite PM200 and PS200, but for specific applications and temperatures it would be possible to formulate other composites to achieve a different optimum balance of properties.

13.6 CONCLUSION

The last two paragraphs may perhaps indicate how we should be thinking in order to solve lubrication problems at high temperatures in future. New and better lubricants will almost certainly be developed, but such developments are likely to be costly and slow. Some radically new lubricants may arise which will solve the problem of providing low friction in air from −100°C to over 1500°C, but for the foreseeable future we cannot rely on that possibility.

In the meantime two new philosophies hinted at above may provide the best hope of progress. The first is to study how we may be able to change a system, or its environment, to suit the best available lubricant. The second is the concept of 'designer' lubricants, optimized to suit a particular application. Each of these, in itself, gives us some prospect of useful progress: jointly they may be really productive.

Even if much better lubricants become available, the synergy of these two concepts will ensure the greatest possible potential for them.

Bibliography

The Tribology Handbook, M. J. Neale (Editor), 1973, Butterworth, London.

Theory and Practice of Lubrication for Engineers, D. D. Fuller, 1984, John Wiley, New York.

Elasto-Hydrodynamic Lubrication, D. Dowson and G. R. Higginson, 1966, Pergamon Press, Oxford.

Synthetic Lubricants, R. C. Gunderson and A. W. Hart, 1962, Reinhold Publishing, New York.

Advanced Bearing Technology (NASA SP-38), E. E. Bisson and W. J. Anderson (Editors), 1954, National Aeronautics and Space Administration, Washington DC.

Interdiscliplinary Approach to Friction and Wear (NASA SP-181), P. M. Ku (Editor), 1968, National Aeronautics and Space Administration, Washington DC.

Interdisciplinary Approach to the Lubrication of Concentrated Contacts (NASA SP-237), P. M. Ku (Editor), 1970, National Aeronautics and Space Administration, Washington DC.

Interdisciplinary Approach to Liquid Lubricant Technology (NASA SP-318), P. M. Ku (Editor), 1973, National Aeronautics and Space Administration, Washington DC.

Modern Lubricating Greases, C. J. Boner, 1976, Scientific Publications, Broseley (UK).

Solid Lubricants and Self-Lubricating Solids, F. J. Clauss, 1972, Academic Press, New York.

Metallic and Ceramic Coatings, M. G. Hocking, V. Vasavtasree, and P. S. Sidky, 1989, Longman Scientific, Harlow.

Gas Film Lubrication, W. A. Gross, 1962, John Wiley, New York.

Gas Lubricated Bearings, N. S. Grassam and J. W. Powell, 1964, Butterworth, London.

Chemical Society Quarterly Reviews, 1971, **25**(2), 265: 'The kinematics of hydrocarbon autoxidation in the liquid phase' (J. Betts).

NASA Technical Memorandum 83465, August 1983: 'Review of liquid lubricant thermal/oxidative degradation' (W. R. Jones).

European Space Research Organisation Contractor Report ESRO CR-402, May 1974, 'Molybdenum disulphide lubrication' (A. R. Lansdown).

Fundamentals of High Temperature Friction and Wear with Emphasis on Solid Lubrication for Heat Engines, F. F. Ling, 1985, Industrial Tribology Institute, New York.

Tribology, 1973, **6**(6), 219: 'Dry bearings: a survey of materials and factors affecting their performance' (J. K. Lancaster).

INDEX

Printed and bound by CPI Group (UK) Ltd, Croydon, CR0 4YY
18/09/2026
14964996-0001